A Volume in
Exploring Complexity

Volume Two
Classic Complexity:
From the Abstract to the Concrete

Exploring Complexity: Volume Two

Classic Complexity:
From the Abstract to the Concrete

Edited by
Kurt A. Richardson
Jeffrey A. Goldstein

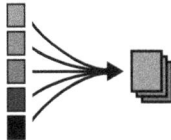

ISCE
Publishing

395 Central Street
Mansfield, MA 02048

Classic Complexity: From the Abstract to the Concrete
Exploring Complexity Book Series: Volume 2
Edited by: Kurt A. Richardson & Jeffrey A. Goldstein

Library of Congress Control Number: 2007926713

ISBN: 0-9791688-3-X
ISBN13: 978-0-9791688-3-3

Copyright © 2007 ISCE Publishing, 395 Central Street, Mansfield, MA 02048, USA

Printed in the United States of America

CONTENTS

Section 3
Complexity and Organization

Volume Editorial:
On the importance of Classic Papers

Kurt A. Richardson
Jeffrey A. Goldstein

With a field so broad in both breadth and depth as that making up the contemporary study of complex systems, it is well nigh impossible to cover all the bases underlying the development of well argued, robust and relevant understanding of these systems. That is one of the reasons *Emergence: Complexity and Organization* (E:CO) has offered a Classical Paper in each issue. Now, in this current volume we have made available in one publication venue the diverse Classical Papers we have published so far. These papers are offered not only to enrich our current understandings by exhibiting the historical background to many of today's leading complexity-based ideas, perspectives, and methods. They are also gathered here to help address some of the difficulties confronting not only complexity thinkers, but for that matter any thinker sincerely trying to grasp the novel situations and novel difficulties we face in modern times. These difficulties include:

Chinese Whispers: As past writings move ever further into the past the author's ideas tend to be retranslated into the language of the time in which they are read. Not only has the intellectual landscape evolved considerably, but so have the social and political landscapes, all of which can act to hinder a 'true' reading of the author's original text. Given the vastness of the material available it is no surprise that many of us only know the works of previous authors through short quotations used by modern writers to support a particular point of view. The Chinese Whisper problem arises because quotes are always taken out of context in order to support the current author's position. Indeed, the American philosopher, sociologist, and proto-complexity thinker George Herbert Mead developed his own prescient ideas on the key complexity notion of emergence through a radical re-assessment of temporality, whereby the past itself undergoes transformation from the point of view of the emerging "Philosophy of the Present." (Mead, 1932). One could put Mead's insight into the quip: "the past is not what it used to be!"

If another author has only had access to the original author through quotations made by more recent authors, it is no surprise that over time the meaning originally intended by the original author slowly shifts - sometimes to the point of not even being reconcilable with the original text. In politics this is often done maliciously, but in science it is an unfortunate by-product of trying to assimilate the overwhelming accumulation of past thoughts and writings (though this also possesses a positive creative dimension). If you really want to know what a long gone author said on a particular subject then you really need to explore their original texts. Even then your modern cultural biases will prevent you from obtaining exactly the original senses (researching the history surrounding the author often allows for a more genuine interpretation - though who has the time to perform such deep research into every author they take an interest in?) - you will at least remove the layers of reinterpretation and reframing imposed by many intermediate interpreters.

Reinventing the Wheel: With all that has gone before it is no surprise whatsoever that the wheel gets reinvented time and time again. Moving in and out of favor is part of the normal evolution of any idea, though each reinvention often changes the previous incarnation in some subtle, and not so subtle, ways. Many of us involved in the *E:CO* project are profoundly interested in philosophical issues along with scientific concerns. It seems the more we read the writings of past philosophers the more modern ideas can be uncovered there. Indeed, it wouldn't surprise us at all if one could put together a modern text on complexity thinking using only Ancient Greek texts! Yet, even with the best intentions in the world, and a never-ending budget, we can never take everything that has gone before into account. Since, very few new ideas are completely original, exploring past texts can reduce the effort expended in searching for novel solutions - but it can also drive the wheel of reinvention as well. Claims of originality in the modern world where there is no basis for such claims seems to be an integral part of the Market-driven economy. However, originality in and of itself may not be as important as one might think - which is rather fortunate given its rareness. An old idea brought to a new audience can itself spark further original ideas. In absolute terms an old idea reframed for modern consumption really is a new idea and so the notion of reinventing the wheel may not be so bad after all.

Empire Building: The development of any school of thought is often associated with a rush to differentiate itself from other schools of thought. This often means highlighting differences and subduing similarities. Empire building in science often leads to 'isms', i.e., a collection of coherent thoughts and theories that seem self-contained, and quite distinguishable from other 'isms'. Many complexity writers begin their treatises by expounding the shortcomings of so-called 'Newtonian thinking' and develop a new way, 'Complexity thinking', built upon corrections to those prior limitations. The education theorist John Dewey noted exactly the same process in the development of progressive educational methods from traditional methods. Progressivism emerged in response to the perceived shortcomings of 'traditionalism', rather than the perceived needs of Education itself. Dewey says:

"*For in spite of itself any movement that thinks and acts in terms of an 'ism becomes so involved in reaction against other 'isms that it is unwittingly controlled by them. For it then forms its principles by reaction against them instead of by a comprehensive, constructive survey of actual needs, problems, and possibilities.*" (Dewey, 1938: 6).

We've actually lost count of the number of times we've read that "linear thinking is dead -long live nonlinear thinking" -a rather pertinent illustration of reactionary-based theory development, which severely limits the possibility of realizing the full implications and extent of complexity.

An example of a literature that seems to have been arbitrarily (not necessarily intentionally) sidelined by the complexity community is that of *soft systems thinking* which evolved from the mathematically-based general systems theory. Offspring literatures include *critical systems heuristics*, *critical systems thinking*, and *boundary critique* to name but a few. Through these different fields GST moved from the abstract world of mathematics (and some might say, un-appliable) to the concrete, though vague, world of useful application. Complexity thinking has only just taken the first steps in the direction towards application. However, in the effort to distinguish complexity thinking from general systems theory and its many outgrowths, complexity thinkers are severely limiting their subject's potential.

Of course *Empire Building*, like all seemingly negative concepts, has its more positive flip-side: it helps a community keep some degree of focus. This is essential if (partial) solutions to emerging problems/issues are to be found. The trick is to not reify one's Empire's boundaries in a politically motivated effort to differentiate and marginalize.

Historicity: Not all papers are created equal. Even though the obscurest of papers will have had some direct influence on the evolution of thought (although complexity places profound limitations on our ability to judge such influence objectively), some papers really do deserve the honor of being called *seminal*. Sometimes entire fields of endeavor can be traced back to a single paper (although many less known papers probably inspired the seminal one) - Einstein's 'special relativity' for example. Understanding the historical arising and development of one's field of interest undoubtedly leads to a richer understanding of that particular field, which can then open up avenues of exploration that might not have previously seemed legitimate.

To be sure, in future volumes of *E:CO* we aim to reprint further classical papers. Each paper will be chosen to exemplify at least one if not all of the issues briefly discussed above and will be introduced by whomever recommended the paper, or who otherwise has some expertise in the area covered by the paper. We also aim to present the papers in their original formatting, rather than re-typesetting them. As such, the printing quality will not be up to our usual standards, but we think it is important to see the papers in precisely the form they first appeared (even if we cannot reproduce to culture backdrop).

Dewey, J. (1938). *Experience & Education*, New York, NY: Touchstone, ISBN 0684838281.

Mead, G. H. (1932). *The Philosophy of the Present: Supplementary Essays*, A. E. Murphy (ed.), La Salle, Illinois: Open Court.

Complexity and Philosophy

1. Emergence
Stephen C. Pepper

Originally published as Pepper, S. C. (1926). "Emergence," *The Journal of Philosophy*, 23(9): 241-245. Reproduced with the kind permission of The Journal of Philosophy.

Emergence then and now: Concepts, criticisms, and rejoinders

"We seem to be in the presence of a perfectly good dilemma: We must either explain things by what they are or else by what they are not. If we explain them by what they are, we leave them unexplained. If we explain them by what they are not, our explanation is fallacious."

William Ernest Hocking (1941)

The evolution of emergence

Although it is not generally well-known among complexity afficionados, the concept of *emergence* had a well established history before the advent of the present day study of complex systems. As early as 1874, the American/British philosopher and man of letters G. H. Lewes (1874-1879) had coined the term "emergent" in its modern technical meaning: "...although each effect is the resultant of its components, we cannot always trace the steps of the process, so as to see in the product the mode of operation of each factor. In the latter case, I propose to call the effect an emergent. It arises out of the combined agencies, but in a form which does not display the agents in action."[1] Several decades later this nascent notion of 'emergent' was elaborated into a process of 'emergence' which became the basis for a loosely joined scientific and philosophical movement called 'Emergent Evolutionism' (for history and review see Blitz, 1992). Such eminent Emergent Evolutionist philosophers and scientists as Samuel Alexander, C. L. Morgan, C. D. Broad, W. Wheeler, A. N. Whitehead, and others discussed emergence in terms of a sudden arising of new 'collocations' or 'integrations' with new properties arising on a new 'higher' emergent level out of 'lower' level components. This enriched concept of emergence was offered as a counter to the then prevalent interpretation of evolution as taking place through incremental steps, a process understood in mechanistic terms. In contrast, Emergent Evolutionists held that a scientific and philosophical perspective founded on emergence was capable of steering between the extremes of *mechanistic reductionism* on the one side and an ungrounded *vitalism* on the other (see Goldstein, 1999, 2003).

Emergent Evolutionism as a movement died-out by the mid-nineteen thirties but the idea at its heart proved to have staying power as it found expression in the philosophy of science, process philosophy and theology shaped by Whiteheadian metaphysics, theoretical biology, and the burgeoning arena of neuroscience. The dominant emphasis continued to be the capacity for the idea of emergence to combine an anti-mechanistist / anti-reductionist stance with a way to talk about higher level organization and its novelty, and doing so without passing over into a supra-naturalism (Goldstein, 2000). These advantages have continued into contemporary research into complex systems where the idea of emergence has moved from mostly armchair speculation into actual laboratories, physical, computational and social. Nowadays when the term 'emergence' is used, it refers to a group of phenomena sharing a family resemblance with the following features (see Goldstein, 1999):

- Macro- or global level in contrast to the micro-level locus of the components;
- Coherence, correlation, or integration at the macro-level;
- Radical novelty of the higher level expressed as its unpredictability and nondeducibility from as well as irreducibility to lower level components;
- Ostensiveness since this radical novelty only shows itself as the system is observed, not ahead of time;
- Dynamical in the sense that emergent phenomena are not pre-given wholes but arise as a complex system evolves over time;

Examples of emergence in complex systems include the coherence seen in various kinds of phase transitions, the new patterns and properties exhibited in so-called self-organizing systems, higher level patterns and structures found in simulations like cellular automata and multi-agent models, collective level behavior arising in networked systems (whether social, technological, or computational), and more recently the 'quantum protectorates' studied in the field known as *complex, adaptive matter*, about which more will be discussed below.

Although embraced by quite a few prestigious thinkers, the Emergent Evolutionist adumbration of the idea of emergence did not go without detractors, chief among whom, during the nineteen twen-

4

ties, was the influential philosopher Stephen Pepper, more widely known later on for his seminal work *World Hypotheses* (Pepper, 1942). *E:CO* is presenting Pepper's article on emergence from 1925 for at least two reasons. First, although Pepper's article was critical of the idea of emergence, his description was fairly faithful to what the early emergentists had in mind. We can accordingly compare that older meaning of emergence with the current one. Second, by taking a close look at Pepper's reductionist argument against emergence, we can appreciate how a commitment to the idea of emergence has generally run counter to reductionist explanatory strategies. I will also go over some of the typical fault lines of such reductionist arguments.

Collapsing new variables into old

In his critique, Pepper took on the emergentist position that new variables were needed in order to represent the new 'higher' level emergent phenomena and their dynamics. What he attempted to demonstrate instead was that any such new variable could in actuality be collapsed into pre-existing variables that were already in use (or least could potentially be used) in representing lower level functional relationships[2]. The first step of Pepper's attack was his contention that the new emergent level must be describable in one of two ways: (1) as a new type of functional relation among the already existing variables of the system; or, (2) as a functional relationship among new variables which are revealed on the new level (see discussion of Pepper's critique in Meehl & Sellars, 1956). In the first case of a new type of functional relationship among old variables, Pepper argued that any new functional relation could always be expressed by some kind of modification of already existing relations, here showing his card as a modern-day Anaxogorian in denying that any modification, rearrangement, or restructuring could possibly introduce genuine novelty. Pepper's argument reflected a persistent assumption of hard core reductionists, namely, that the coming into being of what is radically novel is simply not possible, a pessimistic perspective so eloquently expressed in the Biblical Book of *Ecclesiastes*, "What has been is what will be, and what has been done is what will be done; and there is nothing new under the sun. Is there a thing of which it is said, "See, this is new"? It has been already, in the ages before us."

Regarding the second case, that of the need on the part of emergentists to introduce new variables to represent the emergent level, Pepper attempted to prove how they too (and their functional relationships) were nothing more than an elaboration, no matter how complicated, of a functional relation between old variables. To do so he explicitly assumed, first of all, that emergence was the result of a deterministic process (the only option for him if it was not to be supra-naturalist in origin) and therefore could not expropriate randomness. To be sure, complexity theory has witnessed a repudiation of this wide spread presumption[3]. For Pepper, though, the purportedly new variables involved in deterministic processes must either possess some kind of functional relationship with old variables on the lower level or they do not. In the first option, if the new variables of the emergent level do indeed have a functional relationship to the lower level variables, then, according to Pepper, these new variables must necessarily be expressible in terms of the lower level variables since, for Pepper, the mere existence of a functional relationship implied that one set of variables could be translated to the other given the appropriate means of expressing the new variables in terms of the old. This was really a restatement of Pepper's earlier argument stated in the previous paragraph. He concluded that "…[the new variables would] have to drop down and take their place among the lower level variables as elements of a lower level shift."

Pepper now had to come to terms with the second option, i.e., when the new variables purportedly required by the new emergent phenomena did not have a functional relationship with the old variables. According to Pepper, if the new variables didn't have a functional relationship with the old ones, this implied that the emergent phenomena expressed by these disconnected variables must lack the potency necessary for what the philosophers of science Meehl and Sellars (1956) in their commentary on Pepper's argument called "making a difference." In other words, Pepper concluded that emergent phenomena would amount to nothing more than mere epiphenomena, again, a point of view not uncommonly found among reductionist detractors of emergence.

There were several faulty moves in Pepper's arguments, not the least of which was their pervasive question-begging. For example, Pepper first assumed that any kind of purported genuine novelty attributed to emergent phenomena could always be shown to be an epiphenomenon by definition and thus he simply could not imagine any natural process powerful enough to bring about the radically original. Indeed, we can discern in Pepper's posture towards emergence a manifestation of two linchpins in the then prevailing picture of natural change, viz., first that deterministic processes must abjure the incorporation of chance and, second, they could not possess operations powerful enough to bring about the kind of radical novelty entailed by a doctrine of emergence. Yet, the study of complex systems has shown that the radical novelty of emergent phenomena can be the result of both an appropriation of chance (or 'noise') as well as iterative and combinatory operations. The eminent philosopher of science Karl Popper (quoted in Stephan, 1992: 34-35), who, incidentally, later entertained his own view of emergence, characterized anti-emergentist positions like Pepper's in these words, "Given the precise arrangement of the atoms it should in principle be possible, the argument goes, to derive, or to predict, all the properties of every new arrangement from a knowledge of the 'intrinsic' properties of the atoms," an argument Popper disagreed with by asserting instead it was indeed possible for new arrangements to lead to physical and chemical properties not derivable from like-minded 'atomistic' theories.

Reductionism's preparatory destruction of emergent level phenomena

When Pepper and others of his ilk look to emergent phenomena as epiphenomena, they presume the lack of any real causal efficacy on the part of emergent level phenomena by first imaginatively destroying the integrity of emergent level wholeness. That is, Pepper's approach to emergence, like all totalistic reductionist positions, can only work by ignoring all the preparatory destruction that precedes and accordingly makes possible reductionist explanations, namely, a preliminary destruction of upper emergent level phenomena particularly as regards their wholeness. Indeed, it has been pointed out that much of the ingenuity of reductionist explanations lies precisely in upper level features being destroyed, either directly or indirectly (see Sklar, 1995). After all, 'reduce' means 'to decrease' or 'to diminish'.

The direct type is like what the philosopher Thomas Nickles (cited in Wimsatt, 1994) has discussed as *transformative operations* similar to the reduction of ores to metals or wood to pulp, cases where it is obvious that such higher level features (e.g., the grain of wood) do not survive in the pulp. In a similar vein, the complexity oriented neuro-physiologist Jack Cowan once described the difference between the subject matters of biophysics and theoretical biology in the following vivid manner: take an organism and homogenize it in a Waring blender, the biophysicist is interested in those properties which are invariant under that transformation (cited in Wimsatt, 1997). Because a Waring blender type of destruction will typically eliminate higher level features entirely out of existence, features which need to be accounted for by a credible reducing theory, the philosopher of science William Wimsatt (1974) has cautioned that adequately formulated "bridge" laws included in reductive explanations need to include qualifications concerning the mode and extent of the destructiveness wrought by the reductive explanation.

The indirect type of preparatory destruction contained in reductionist explanations has to do with the stripping away of higher level qualities that has already taken place as part of the general scientific climate. An example provided by the philosopher Robert Nozick (1981) concerns how Maxwell's brilliant, but reductive identification of light with electro-magnetism only, could have been achieved after science had first stripped light of its sensed qualities of color and hue[4]. According to Nozick, one of the main reasons reductionist explanations can appear compelling at all is their appeal to theoretical primitives which do not possess the qualities of the phenomena under question and thus impart the sense of going "deeper." This is related to the quote from Hocking above who presented the paradox of explaining something by what it is not. For instance, the thermodynamic property of heat is explained by heat-less statistical mechanics and Mendelian laws by gene-less Mendelian properties. The theoretical primitives of reductionist explanations must not be allowed to possess higher level properties. In fact, as Nozick has remarked, if the lower level, reducing theory contained the exact same properties as the upper level, reduced theory, the explanation might get caught in an endless loop of self-reference, upper level property referring to lower level property referring to upper level property, around and around in a vicious circle. We can conclude, consequently, that one of the ways reductionist explanations operate is to stop this vicious circle by a preliminary making sure that the upper level is robbed of its possible *sui generis* character before the explanation can even proceed. And once rid of its valence, the higher emergent level no longer poses a particular challenge since the reductionist does not how have to imagine the possibility of processes or operations that could bring about this higher level valence.

This contention that too much is destroyed by reductive explanations is at the heart of the Nobel Laureate, condensed matter physicist Philip Anderson's (1972) famous anti-reductionist and pro-emergentist *Constructionist Hypothesis* which states that the ability to reduce everything to simple fundamental laws does not then imply the ability to start from those laws and reconstruct the universe. Humpty Dumpty, alas, once he's had his great fall from his higher level perch and is then broken into the many shards of his lower level components cannot be put back together again! For Anderson, what ineluctably gets in the way of the possibility of 'reconstructing' the universe from reduced fragments and their simple laws are both scale and complexity, since at each new level of complexity entirely new properties appear requiring new laws, constructs, and generalizations. Consequently, Anderson also emphasized the need for a hierarchy of sciences, an idea repeatedly appealed to on the part of emergentists, each science focusing its theoretical and empirical energies on a specific level, and with the methods, constructs, and theories of each science not reducible to the one beneath. Indeed, Anderson's point can be understood as an echo almost a century later of Lewes' definition whereby an "emergent," unlike a "resultant," cannot always be traced back, "... so as to see in the product the mode of operation of each factor."

The constructive destruction of the lower level in emergence

From a different angle, another kind of destruction can be discerned in the process of emergence itself. That is, one of the reasons why emergent phenomena are not essentially amenable to reduction is that the processes involved in emergence possess a certain kind of destruction of lower level elements with the result that these lower level elements are no longer there, at least in the manner they previously existed, to be reduced to! This is quite different than the intentional preparatory destruction on the part of reductionists for the destructive element in emergence is in fact a necessary condition for emergence to have the capacity to lead to unpredictable, nondeducible, and unpredictable outcomes. In other words, the higher emergent level cannot be reduced to the lower level since that the entities and properties on the lower level have in effect been destroyed in the constructional building up of emergent order.

Such a scenario for emergence was obliquely noticed by one of the chief philosophical critics of early emergentism, Charles Baylis (see Blitz, 1992). Baylis argued that one aspect of emergence would involve the destruction of those properties the parts had before they were emergently combined, referring here explicitly to the way properties are transformed in chemical reactions, e.g., the gaseous nature of hydrogen and oxygen being destroyed when they are fused into water with its property of liquidity. More recently, the philosopher Paul Humphries (1997) has pointed to just such a destruction in his characterization of emergence in terms of a "fusion," the latter a catch-all term for processes leading to emergent outcomes. It is precisely this destructive side of emergence that Humphries appeals to in his suggestion that reduction doesn't make sense as a strategy for explaining emergents precisely because the lower level to which the explanation is supposed to be reduced to has effectively been destroyed during the processes of fusion. In this sense of destruction accompanying emergence, we can say the whole is less than the sum of the parts - the sum of the parts of the oxygen and hydrogen atoms not in the form of H_2O would include properties that are not found when they are combined to form H_2O. It is this destructive aspect of emergence which can help account for what Lewes referred to as untraceability of emergents in contrast to resultants.

Emergence and new variables in complex systems

In contrast to Pepper's arguments against the viability of new variables being introduced to explain higher level emergent phenomena, order parameters are now commonly used along with a control parameters to focus the attention on the higher level order arising during phase transitions. For instance, in an unmagnetized state of iron, the atomic spins, not having a preferred orientation, point in all different directions; a state described as possessing a high degree of symmetry since at any particular location in the system what is happening on either side is roughly the same as in a mirror image. At a comparatively higher temperature (the metric of which is the control parameter), this high symmetry is associated with disorderly motion occasioned by the affect of fluctuations. However, by reducing the temperature, the spins become aligned in one direction thereby breaking the initial symmetry and leading to the commencement of magnetization due to the newly arisen preferred directionality of the spins.

Another example of symmetry breaking and the accompanying emergence of new order can be found in the phase transition of a nematic liquid crystal which, at first, is disordered with rotational symmetry but, undergoing a change of certain thermodynamic parameters, becomes ordered characterized by a breaking of rotational symmetry towards a special direction (Anderson & Stein, 1987). An order parameter therefore represents emergent order coincident with the breaking of symmetry. Its value would be zero at the symmetrical, disorderly phase and one in a totally ordered phase.

The synergetics approach founded by the eminent German physicist Hermann Haken (1981, 1987) has demonstrated that large collectives, whether complex materials, societies or brains, may be analyzed globally with the use of order parameters. Local interactions among the individual members of the collective result in the emergence of long term correlations in the behavior of these individuals. These correlations in turn give rise to order parameters, or macroscopic variables, which describe the global behavior of the collective. These order parameters are frequently 'slaved' to other control parameters which may be externally modified to influence the global behavior of the collective.

The idea of higher level variables such as order parameters also opens the way towards findings of universalities among different systems which can be understood as indications of emergence since they represent new laws. The physicist and Nobel Laureate Robert Laughlin (2005) has recently put forward the strong thesis many scientific laws are in general emergent since they involve higher level organizing principles. Van Gelder (cited in Clark, 1996) has suggested that the variables expressed in dynamical explanations are not about the underlying dynamics anyway, but about the dynamical region which is understood in terms of the global patterns showing up in phase space portraits and other 'qualitative' dynamics. Moreover, Clark cites Luc Steels's distinction between controlled and uncontrolled variables. Whereas the first refers to variables that can be directly controlled, e.g., a robot can directly increase or decrease its speed, the second changes indirectly as a side-effect. Clark (1996) thereby defines emergence as involving those phenomena whose "roots involve uncontrolled variables and are thus the products of collective activity rather than dedicated components or control systems" (p. 267). Furthermore, the real locus of the uncontrolled variable is the relation between the system and its environment. In this way, the variables can themselves be emergent.

A resurgence of emergence and higher level organizing principles

Recently, a group of condensed matter physicists making-up the Institute for Complex Adaptive Matter have been using the term 'emergent' explicitly to describe scientific laws pertaining to higher level organizing principles. One of the members of this group, the aforementioned Robert Laughlin (2002) cogently argues that scientific laws in general are emergent in the sense that they represent higher level organizing principles, are "collective in nature," are "encoded only indirectly by the underlying laws of quantum mechanics, and in a deep sense independent of them," and "are exact only in the thermodynamic limit." Writing with his colleague David Pines, Laughlin (Laughlin & Pines, 2000) further writes:

"The emergent physical phenomena regulated by higher organizing principles have a property, namely their insensitivity to microscopics that is directly relevant to the broad question of what is knowable in

the deepest sense of the term. The low energy excitation spectrum of a conventional superconductor, for example, is completely generic and is characterized by a handful of parameters that may be determined experimentally but cannot, in general, be computed from first principles. An even more trivial example is the low-energy excitation spectrum of a conventional crystalline insulator, which consists of transverse and longitudinal sound and nothing else, regardless of details. (p. 29)

Other candidates for emergence offered by Laughlin and Pines include so-called "quantum protectorates" like the crystalline state, the fractional quantum Hall effect and "quasiparticles." Laughlin (2002, 2003) has gone even further suggesting that relativity, renormalizability, gauge forces, fractional quantum numbers, and the Big Bang itself are all genuinely emergent phenomena.

Conclusion

Although as we've seen above, the idea of emergence has gone through signficant changes during its century and a quarter life span, certain crucial features have stayed put, namely, its focus on higher level organizing principles as the key to explanation as well as understanding, and its being as alternative to pure reductionism. In respect to its use in scientific explanation, the construct of emergence is appealed to when the dynamics of a system seem better understood by focusing on across-system organization rather than on the parts or properties of parts alone. But here emergence functions not so much as an explanation but rather as a descriptive term pointing to the patterns, structures, or properties that are exhibited on the macro-level (see Goldstein, 1999).

This emphasis on the descriptive role of emergence is by no means to be taken as downplaying the importance of the idea of emergence, but is offered rather to highlight how emergence functions more as the starting point rather than the terminus of an explanation. As a result, recognizing the emergent level of patterns, dynamics, and properties is when deeper exploration of the laws governing complex systems can begin. It is to further the appreciation of the critical moment of recognizing this emergent level that Pepper's article on emergence is being offered here in the pages of *E:CO*.

Jeffrey A. Goldstein

References

Anderson, P. (1972). "More is different: Broken symmetry and the nature of the hierarchical structure of science," *Science*, 177 (4047): 393-396.

Anderson, P. and Stein, D. (1987). "Broken symmetry, emergent properties, dissipative structures, life: Are they related?" in F. E. Yates, A.Garfinkel, D. Walter and G. Yates (eds.), *Self-organizing systems: The emergence of order (Life science monographs)*, NY: Plenum Press, pp. 445-457.

Blitz, D. (1992). *Emergent evolution: Qualitative novelty and the levels of reality*, Dordrecht: Kluwer Academic Publishers.

Bunge, M. (1979). *Causality and modern science*, NY: Dover.

Clark, A. (1996). "Happy couplings: Emergence and explanatory interlock," in M. Boden (ed.), *The philosophy of artificial life*, Oxford, England: Oxford University Press, pp. 262-281.

Goldstein, J. (1999). "Emergence as a construct: History and issues," *Emergence*, 1(1): 49 72.

Goldstein, J. (2000). "Emergence: A construct amid a thicket of conceptual snares," *Emergence*, (2)1: 5-22.

Goldstein, J. (2003). "The construction of emergent order, or how to resist the temptation of hylozoism," *Nonlinear Dynamics, Psychology, and Life Sciences*, 7(4): 295-314.

Haken, H. (1981). *The science of structure: Synergetics*, F. Bradley (trans.), NY: Van Nostrand Reinhold Company

Haken, H. (1987). "Synergetics: An approach to self-organization," in F. E. Yates, A. Garfinkel, D. Walter and G. Yates (eds.), *Self-organizing systems: The emergence of order (Life science monographs)*, NY: Plenum Press, pp. 417-433.

Hocking, W. E. (1941). "Whitehead on mind and nature," in P. Schilpp (ed.), *The philosophy of Alfred North Whitehead*, Carbondale, Illinois: Library of Living Philosophers, pp. 381-404. .

Humphreys, P. (1997). "How properties emerge," *Philosophy of Science*, 64: 1-17.

Laughlin, R. B. (2002). "The physical basis of computability," *Computing in Science and Engineering*, 4 (27). Online version.

Laughlin, R. B. (2003). "Emergent relativity," in *Frontiers in science: In celebration of the 80th Birthday of C. N. Yang*, Singapore: World Scientific. Online version.

Laughlin, R. B. (2005). *A different universe: Reinventing physics from the bottom down*, NY: Basic Books.

Laughlin, R. B. and Pines, D. (2000). "The theory of everything," *Proceedings of the National Academy of Sciences*, (97)1: 28-31. Online version.

Lewes, G. H. (1874-1879). *Problems of life and mind*, London: Truebner.

Meehl, P. and Sellars, W. (1956). "The concept of emergence," in H. Feigl and M. Scriven (eds.), *The foundations of science and the concepts of psychology and psychoanalysis. (Minnesota studies in the philosophy of science, Vol. 1)*. Minneapolis: University of Minnesota Press, pp. 239-252.

Nozick, R. (1981). *Philosophical explanations*, Cambridge, MA: Harvard University Press.

Pepper, S. (1942). *World hypotheses: A study in evidence*, Los Angeles: University of California Press.

Stephan, A. (1992). "Emergence: A systematic view on its historical aspects," in A. Beckermann, H. Flohr and J. Kim (eds.), *Emergence or reduction: Essays on the prospects of nonreductive physicalism*, Berlin: Walter de Gruyter, pp. 25-47.

Sklar, L. (1995). *Physics and chance: Philosophical issues in the foundations of statistical mechanics*, Cambridge, England: Cambridge University Press.

Wimsatt, W. C. (1994). "Levels of organization, perspectives, and causal thickets,"

Canadian Journal of Philosophy, Supp. 20: 207-274.

Wimsatt, W. C. (1974). "Complexity and organization," in K. F. Schaffner and R. S. Cohen (eds.), *PSA 1972, Proceedings of the Philosophy of Science Association*, Dordrecht: Reidel, pp. 67-86.

Notes

[1] Lewes went on: "Every resultant is either a sum or a difference of the co-operant forces ... every resultant is clearly traceable in its components because these are homogenous and commensurable ... the emergent is unlike its components in so far as these are incommensurable, and it cannot be reduced either to their sum or their difference..." (pp. 368-369).

[2] Pepper was here appealing to the accepted use of mathematics to express functional relationships within a system (see Bunge, 1979; Bunge also pointed out that emergent laws need not be entirely new or absolutely new, but merely have to be new in regard to the laws followed by the object under question).

[3] The exclusion of chance from explanations involving natural processes was prevalent at Pepper's time despite the critical role of chance in evolution as well as Charles Sanders Peirce's notion of *tychism* which rather presciently foreshadowed the recent reconciliation of determinism and apparent randomness in the case of chaos theory.

[4] Indeed, reductionism since Galileo has relied upon a previous stripping away of so-called 'secondary' qualities from phenomena so as to not be encumbered by them in explanation. Thus, as Sklar has suggested, Maxwell didn't say the secondary quality of color was somehow correlated with electro-magnetic waves, rather color was dismissed as non-essential to his theory of electromagnetism.

EMERGENCE

E MERGENCE signifies a kind of change. There seem to be three important kinds of change considered possible in modern metaphysical discussion. First, there is chance occurrence, the assertion of a cosmic irregularity, an occurrence about which no law could be stated. Second, there is what we may call a "shift," a change in which one characteristic replaces another, the sort of change traditionally described as invariable succession and when more refined described as a functional relation. Thirdly, there is emergence, which is a cumulative change, a change in which certain characteristics supervene upon other characteristics, these characteristics being adequate to explain the occurrence on their level. The important points here are first, that in discussing emergence we are not discussing the possibility of cosmic chance. The emergent evolutionists admit a thorough-going regularity in nature. And secondly, we are not discussing the legitimacy of shifts. These also are admitted. The issue is whether in addition to shifts there are emergent changes.

The theory of emergence involves three propositions: (1) that there are levels of existence defined in terms of degrees of integration; (2) that there are marks which distinguish these levels from one another over and above the degrees of integration; (3) that it is impossible to deduce the marks of a higher level from those of a lower level, and perhaps also (though this is not clear) impossible to deduce marks of a lower level from those of a higher. The first proposition, that there are degrees of integration in nature, is not controversial. The specific issue arises from the second and third propositions. The second states that there is cumulative change, the third that such change is not predictable.

What I wish to show is that each of these propositions is subject to a dilemma: (1) either the alleged emergent change is not cumulative or it is epiphenomenal; (2) either the alleged emergent change is predictable like any physical change, or it is epiphenomenal. I assume that a theory of wholesale epiphenomenalism is metaphysically unsatisfactory. I feel the more justified in making this assumption because I have been led to understand that the theory of emergent evolution has been largely developed as a corrective of mechanistic theories with their attendant psycho-physical dualisms and epiphenomenalisms.

Let us now exhibit the first dilemma. There seem to be two theories as to the nature of the marks which emerge and distinguish levels of integration. One theory states that these are qualities like sensory qualities, metaphysical simples. Alexander's system contains a theory of emergent qualities. The term "quality" is ambiguous and frequently made to cover both qualities like Alexander's and laws of activity. Bearing this ambiguity in mind, one will find that most emergent evolutionists have theories of emergent laws—emergent laws being the other sort of marks thought of as emerging.

Now, a theory of emergent qualities is palpably a theory of epiphenomena. It is not so obvious that a theory of emergent laws must also be such—or else cease to be a theory of emergence.

Accurately speaking, we must first observe, laws can not emerge. Emergence is supposed to be a cosmic affair and laws are descriptions. What emerge are not laws, but what the laws describe. This distinction is important, for I maintain that the same natural regularities can be described by a whole hierarchy of different laws and that this ladder of laws has been mistaken by the emergent evolutionists for a ladder of cosmic regularities. But I do not care to develop this matter here.

What I wish to show is that all natural regularities are shifts and can not be otherwise described. Let us suppose a shift at level B is described as a function of four variables q, r, s, and t. Let us then suppose that r and s constitute an integration giving rise to level C at which level a new cosmic regularity emerges that can be described as a function of four variables r, s, a, and b. r and s must necessarily be variables in this emergent law even though they are variables of level B, because they constitute part of the conditions under which the emergent law is possible. Theoretically, to be sure, the emergent law may be thought of either as a function of new variables or as a new function of C-level variables. But actually only the former is possible. For if the new law were not $f_1 (q, r, s, t)$, but were $f_2 (q, r, s, t)$, then, of course, it would *never* be $f_1 (q, r, s, t)$, unless the event were a chance occurrence in which case no regularity could be described anyway. The point is, either f_1 adequately describes the interrelationships of (q, r, s, t) or f_2 does; or if neither adequately describes the interrelationships there is some f_3 that does, but there can not be two adequate descriptions of the same interrelationships among the same variables.

An emergent law must, therefore, involve the emergence of new variables. But these new variables either have some functional relationship with the rest of the lower level variables or they haven't. If they haven't, they are sheer epiphenomena, and the view resolves itself into a theory of qualitative emergence. If they have, they have to be included among the total set of variables described by the lower level functional relation; they have to drop down and take

16

their place among the lower level variables as elements in a lower level shift.

Such being the case, our dilemma is established so far as concerns cumulative change—either there is no such thing or it is epiphenomenal. Now for predictability.

The assumption in theories of emergence is that mechanical changes, single level changes (call them shifts) are predictable, but that emergent changes are not. Now, it must be pointed out at once that in an ultimate sense, no changes of any kind are predictable. One event can never be deduced from another event. Even the first law of motion is not self-evident. From the state of rest or motion of an unmolested body at one moment, you can not deduce the state of rest or motion of that body at the next moment. And what will be the states of two bodies that do molest each other is equally not deducible from their states previous to molestation. Deduction from the states of bodies under certain conditions to their states under other conditions is possible only through the mediation of a descriptive law summarizing a set of shifts in which these bodies and their conditions are variables. It follows that events and qualities occurring on the same level are no more deducible from one another than if they occur on different levels. Cosmic events don't deduce or predict one another; they occur. It is only we who describe them by laws, who also make predictions concerning them by means of our laws. Cosmically speaking, nothing is deducible, and hypothetical emergent qualities or events would be no more peculiar in this respect than the qualities and events that occur on the bed rock level.

It is only humanly speaking that anything is deducible. And what are strictly deducible are neither qualities nor events, but laws. There is a loose sense in which qualities and events are predictable. Thus by a process of interpolation we can and have predicted the properties of missing elements in the periodic table. And by a process of analogy we predict the states of consciousness of other people and of animals. But such loose predictions are not the sort meant by emergent evolutionists. For obviously emergent events and qualities could be predicted in this manner if there were any occasion to predict them. Just such a prediction Alexander makes about God. What the emergent evolutionists mean is that an emergent law could not be deduced from lower level laws as a mathematical theorem can be deduced from other theorems. The condition under which such deduction is strictly possible is that by substitution of one sort or another, one law may be derived from another law or set of laws. It is a natural ideal of science to derive all laws from a certain limited number of primitive laws or principles—not necessarily from one single law—and so to convert science into a mathematics. If it could be assumed that there are no chance occurrences such a system of laws should be obtainable, though it might look very different from the traditional mechanics. The assumption of science appears to be that such a system is obtainable. I do not know what else the dissatisfaction of science with inconsistencies could mean.

Now, there seems to be no intention on the part of emergent evolutionists to deny that such a system is possible or to assert that there are chance occurrences. If that is so, they seem to be faced with the following dilemma: either the emergent laws they are arguing for are ineffectual and epiphenomenal, or they are effectual and capable of being absorbed into the physical system. But apparently they want their laws to be both effectual and at the same time not part of the physical system. Professor Lovejoy, for instance, contrasts laws of behavior with physico-chemical laws. The laws of behavior imply that "some movements of certain complexes of protons and electrons are, in part, functions of other movements which are to occur afterwards. And this determination of some character of a present motion by some character of a yet unrealized future is a conception obviously foreign to the laws of physics and chemistry . . . and it is, in fact, incongruous with the most fundamental methodological assumptions of those sciences." [1] But at the same time he affirms that there is an "assumption that intelligence is capable of being a factor in the control of human actions and thereby of man's physical environment. . . . If *that* postulate, at least, be not true, universities are absurdities and laboratories are costly monuments of delusion" (p. 208). Now clearly, if these laws of behavior are going to step down out of an epiphenomenal heaven and "be a factor in the control . . . of man's physical environment" they are bound soon to get into conflict with physico-chemical laws as now described. I am here, of course, not interested in whether the present movements of electrons and protons are functions of future electrons or protons, or not. But if they are, that important relation would have to be stated among the primitive laws of the physical system or would have to be deducible from some of those laws. That is to say, if the laws of behavior enter into the physical system at any point they must constitute either primitive laws in that system or be deducible from the primitive laws. There is no other way out of it. These supposed emergent laws are either epiphenomena or physical laws. For if they represent irreconcilable inconsistencies in the physical system, they are not laws at all, but statements of chance occurrences.

[1] *Essays in Metaphysics;* Univ. of Cal. Publications, V. 5, pp. 208–209.

So, regarding the characteristics of unpredictability, the situation is the same as regarding the emergent marks; it is a possible characteristic, but only as an epiphenomenal one.

STEPHEN C. PEPPER.

UNIVERSITY OF CALIFORNIA.

2. Novelty, indeterminism, and emergence

W. T. Stace

Originally published as Stace, W. T. (1939). "Novelty, Indeterminism, and Emergence," *The Philosophical Review*, ISSN 0031-8108,
48(3): 296-310. The *E:CO* editorial team would like to thank Duke
University Press, especially Thomas Robinson, for their kind permission to reprint this article.

Background

The Classical Paper of this issue was written in 1939 by the
philosopher W. T. Stace who argued against the metaphysical
viability of emergence in a manner similar to that found in two
previous reprints *in E:CO*, namely, that of Stephen Pepper (1926,
2004) and Charles Baylis (1929, 2006). Stace, born in London and
educated in Scotland and Ireland, spent his early years, 1910-1932,
in the civil service in Sri Lanka (Ceylon), becoming mayor of its main
city, Colombo, at the end of this period. In 1932, Stace's official role
of philosopher started when moved to the US in order to become a
member of the Department of Philosophy at Princeton University
where he stayed until his retirement in 1955. Stace's philosophical writings covered a variety of topics including Hegel, aesthetics,
mysticism, and the free will/determinism controversy. He was a well
regarded philosopher who authored numerous books and articles. In
particular, he championed a view of philosophical 'compatibilism'
which attempted to reconcile strict determinism with free will (Stace,
1952). This is relevant to his attitude toward emergence which we'll
come back to later.

Incidentally, Stace retired to Laguna Beach, California when it
was becoming a center of the counter culture, eventually becoming the
home of the Brotherhood of Eternal Love, the Timothy Leary inspired
commune and source of the infamous 'Orange Sunshine' brand of LSD
(the area used to be an important center for orange groves). I mention
this because in his later years Stace wrote about mysticism, including
the use of psychedelic drugs to induce mystical experiences. As far as
I can tell, Stace didn't fall under the spell of Leary (by the way, another
frequent visitor to Laguna Beach in those years was Charles Manson!).
But not a few notable British intellectuals spent their waning years
waxing high on psychedelics and contemplating the blue Pacific, e.g.,

Aldous Huxley, Gregory Bateson, and others. Thus, I find it interesting that Stace (1960) concluded,

"[I]t is sometimes asserted that mystical experiences can be induced by drugs, such as mescalin, lysergic acid, etc. On the other hand, those who have achieved mystical states as a result of long and arduous spiritual exercises, fasting and prayer, or great moral efforts, possibly spread over many years, are inclined to deny that a drug can induce a 'genuine' mystical experience, or at least look askance at such practices and such a claim. Our principle says that if the phenomenological descriptions of the two experiences are indistinguishable, so far as can be ascertained, then it cannot be denied that if one is a genuine mystical experience the other is also. This will follow notwithstanding the lowly antecedents of one of them, and in spite of the understandable annoyance of an ascetic, a saint, or a spiritual hero, who is told that his careless and worldly neighbor, who never did anything to deserve it, has attained to mystical consciousness by swallowing a pill" (pp. 29-30).

Arguments against emergence

Even though Stace's arguments were critical of the idea of emergence, we can take some of his points, just as we did in the cases of Pepper (Goldstein, 2004) and Baylis (Goldstein, 2006), as indicators of conceptual hurdles that emergence must pass if it is to amount to a viable possibility for nature. In this regard, Stace's paper is instructive for revealing not untypical dispositions towards emergence and its relation to the determinism issue as well as the subjective nature of the recognition of emergent novelty.

Stace's arguments were directed at those philosophical uses of the idea of emergence found in both Emergent Evolutionism, which flourished until the end of the nineteen thirties, as well as the work of Henri Bergson and William James, which had a large influence on the development of the idea of emergence. In particular, Stace disputed the claim that emergent novelty amounted to anything of ontological significance. In a manner calling to mind Baylis's (2006) contention about the arbitrariness of emergent novelty, Stace tried to demonstrate how the recognition of emergent novelty in each case was an arbitrary assignment dependent on the subjective interests of the recognizers.

Moreover, Stace held that the emphasis on novelty *per se* among emergentists arose out of their revolt against the hegemony of science. In that scientific inquiry aimed at explanations expressed in terms of scientific laws (regularities) that came with a version of strict causal determinism, these emergentist thinkers held that science must disallow the emergence of genuine novelty. Furthermore, because the emergence of novelty had to thereby transgress scientific determinism, the emergence of novelty was linked by these thinkers to *in*determinism and to organicity as such, the latter also thought to transgress strict determinism. Such an attitude placing emergence in contrast to science also showed up in the purported antagonism between living organisms and machines. It is worthwhile to point out that a similar attitude about life in contrast to determinism outlived proto-emergentism, finding expression, e.g., in such later ideas as Rosen's (2005) "M/R systems" and Marturana and Varela's (1991) notion of autopoiesis.

It should be noted that the purported enmity between life and machines was questioned as far back as early emergentist thinking and these questions have continued on up to our day. Thus, such diverse thinkers as the proto-emergentist philosopher Samuel Alexander (2004), the mathematicians and computer pioneers Alan Turing (Hodges, 1983) and John von Neumann (2002), and the contemporary artificial intelligence researcher Douglas Hofstadter (1996) have all speculated that over a certain threshold of complexity, 'machines' can behave in ways indistinguishable from the behavior of living organisms. Of course, in this regard we can also consider all the different manifestations of artificial life whose very phrase suggests an overcoming of the earlier stark antagonism between the living and the mechanical.

Stace also challenged the claim that emergentist novelty would necessarily be of ameliorative value, so that a world wherein emergence of novelty takes place would be a better world. The article by David Bella in this issue of *E:CO* also belies such an understanding of emergence.

Determinism/indeterminism

As mentioned above, Stace was committed to upholding a strictly deterministic world view, his doctrine of 'compatibilism' seeking to allow for the co-existence of this commitment along with free will. To accomplish this reconciliation between two apparently contrary philosophical positions, Stace had recourse to a distinction between internal and external 'forces' behind an action. If an action's immediate causes were *internal* psychological states of agent, that is, instigated by a person's thoughts, wishes, emotions, or desires, then it was free. But if, instead, the immediate causes of an action were 'states of affairs' that were *external* to an agent, it was not free. We can see that Stace's 'resolution' of this age old dilemma was based on positing a free arena where determinism didn't reign, namely, the inner world of psychological states and the actions prompted by these states. However, when it came to the world outside of our subjectivity, Stace held to a strict determinism, a presupposition that showed up in his criticism of how some emergentists took up an indeterminist perspective in order to account for emergent novelty in the *external* world.

Stace understood determinism in the customary terms of a predictable causal chain connecting two events, A and B, so that whenever A took place, then B happened and nothing else. In contrast, indeterminism allowed for A to be followed sometimes by B, sometimes by C, sometimes by D, and so on. In such an indeterministic scheme, no conditions existed which could absolutely predict which would follow A, whether B, C, D, or anything at all. In a nod to quantum mechanics, Stace admitted that indeterminism was supposed to rule subatomic realms where it was unpredictable whether an electron would jump to the right or the left (a rather lame example of quantum indeterminacy). But even here, Stace held that an electron's indeterministic unpredictability as to whether it would go left or right, was not really the kind of novel event that emergentists wanted since electrons jumping in that way have been going on for a long time. For Stace, what emergentists really got with indeterminism was the unexpected and not the novel.

Furthermore, Stace pointed out that novelty could indeed take place according to a strictly deterministic viewpoint. For example, an element, A*, from a star, A, not found in our solar system, might in some future scenario combine with a different element, B*, from our sun, B, resulting in a novel combination even though deterministic laws presumably govern the natural processes that led to element A*, element B*, and their combination <A*&B*>.

Repression of the random

Using Stace's own schemata of events A, B, C,..., it needs to be pointed out that quantum indeterminacy does not imply that whatever follows A is altogether unpredictable. Instead, it posits a rather constrained unpredictability having to do with the Uncertainty Principle, which has it that there is an incompleteness involved in the description of a quantum level physical system. This indeterminacy is different from ubiquitous errors in measurements accompanying any experiment; errors which can be corrected for by means of statistics, or improvements in measuring devices. Instead, quantum indeterminacy has it that instead of being understood as a determinate state, a physical system could only be fundamentally characterized by a probability distribution. It was Einstein's uneasiness with this view which prompted his famous remark, "God does not play dice with the universe." But the only way 'dice' would play a role in Stace's characterization would be the direction of the jumping electron, to the left or to the right.

What, therefore, seems to be a bigger issue for Stace was the very rationality of the notion of indeterminism as such. Indeed, this was the position adopted by the Classic Paper of Pepper (2004) who held that a rational explanation had to eschew any sense of indeterminacy. As I've discussed elsewhere (Goldstein, 1997), there is a decided strain in Western thought characterized by not only a dislike and distrust of randomness in rational thought, but even a revulsion towards it, what I've called a "repression of the random." It was in the context of this attitude that Hegel once declared, "Philosophical reflection has no other object than to get rid of what is accidental" (Hegel quoted in Marquard, 1991: 109).

Of course, much has changed in science over the last one hundred years to revolutionize our ideas of causality, particular with regard to the role of indeterminism in quantum mechanics as well as an analogous role of randomness in the study of complex systems. In the study of complex systems, indeterminism enters the picture in at least two places: one is the inclusion of randomization in explanations such as how in simple, self-organizing physical systems, certain features of the ensuing emergent order arises from an amplification of fluctuations; the other is the inclusion of randomization operations which generate complex phenomena. Thus, in the latter, various randomization procedures are explicitly introduced during the generation process in order to bring about novelty, along with other novelty generators such as recombination strategies. Moreover, such phenomena as property of the sensitive dependence on initial conditions of chaotic systems, has taught us that intractable nonlinearities can look surprisingly like indeterminism even when they're not. For these reasons and more, complexity sciences are bringing about a radical reframing of the relation between determinism and indeterminism in our explanations of complex systems.

If emergence only amounted to what Stace (and Baylis and Pepper before him) thought it did, it would not have generated the interest it did then and continues to generate now. I think it is fair to say that, in an important sense, Stace was addressing an idea of emergence whose time had not yet come. At that time it couldn't yet rise up above its armchair status. But today we have experimental methods, methodological procedures, mathematical models, and scientific constructs that much more aptly call for philosophical reflection.

Jeffrey A. Goldstein

References

Alexander, S. (2004). *Space, Time And Deity: The Gifford Lectures At Glasgow*, Whitefish, MT: Kessinger Publishing, ISBN 0766187012.

Baylis, C. A. (1929). "The philosophical functions of emergence," *The Philosophical Review*, ISSN 0031-8108, 38(4):372-384, reprinted in *Emergence: Complexity and Organization*, 8(1): 71-83.

Goldstein, J. (1997). "Embracing the random in the self-organizing Psyche," *Nonlinear Dynamics, Psychology, and Life Sciences*, 1(3): 181-202.

Goldstein, J. (2004). "Emergence then and now: Concepts, criticisms, and rejoinders: Introduction to Pepper's 'Emergence,'" *Emergence: Complexity and Organization*, 6(4): 66-71.

Goldstein, J. (2006). "Introduction to Baylis's Article," *Emergence: Complexity and Organization*, 8(1): 67-70.

Hodges, A. (1983). *Alan Turing: The Enigma*, New York, NY: Simon & Schuster, ISBN 0671492071.

Hofstadter, D. (1996). *Metamagical Themas: Questing for the Essence of Mind and Pattern*, New York, NY: HarperCollins, ISBN 0465045669.

Marquard, O. (1991). *In Defense of the Accidental: Philosophical Studies*, R. M. Wallace (trans.), Oxford, England: Oxford University Press, ISBN 0195072529.

Maturana, H. and Varela, F. (1991). *Autopoiesis and Cognition: The Realization of the Living*, New York, NY: Springer, ISBN 9027710163.

Pepper, S. (1926). "Emergence," *The Journal of Philosophy*, 23(9): 241-245. Reprinted in *Emergence: Complexity and Organization*, 6(4): 72-76.

Rosen, R. (2005). *Life Itself: A Comprehensive Inquiry into the Nature, Origin, and Fabrication of Life*, New York, NY: Columbia University Press, ISBN: 0231075650.

Stace, W. T. (1952). *Religion and the Modern Mind*, New York, NY: Lippincott.

Stace, W. T. (1960). *Mysticism and Philosophy*, ISBN: 0333082745, Philadelphia, PA: J. B. Lippincott.

Von Neumann, J. (2002). *Theory of Self-Reproducing Automata*, A. W. Burks (ed.), Champagne, Illinois: University Illinois, ISBN 0598377980.

NOVELTY, INDETERMINISM, AND EMERGENCE

THE purpose of this paper is entirely analytical. It is not my aim to put forward any positive or constructive thesis. My object is, if possible, to introduce some measure of clarity into what seems to me to be a fog of vague ideas connected with the concept of novelty as that concept is found in the writings of such authors as Bergson, William James, Samuel Alexander, and other contemporary philosophers. I simply want to try to help in clearing up what seems to me to be a sphere of very confused thinking.

Before entering on the detail of this I want to make three introductory remarks. First, the concept of novelty in contemporary philosophy (except perhaps in Alexander) is part and parcel of a philosophical revolt against the overweening pretensions of science. Science finds, or used to find, the world completely governed by law. All events are reduced to cases of causal or functional determination. This means, it is alleged, that there can be no genuine novelty in the world. We shall have a mechanical universe, an eternal repetition of unalterable sequences, the everlasting turning of wheels upon wheels. All change is mere rearrangement of old elements in new patterns. The end is foreseeable in the beginning, is contained in the beginning. The universe cannot produce anything which was not implicitly present from the very beginning, that is, it cannot produce any novelty. The concept of novelty in philosophy is a revolt against this mechanical view of the world which is the product of science.

Secondly, it is a revolt based upon an *emotional* revulsion. The philosophers of novelty *dislike* the scientific picture. They *desire* a world in which what they call genuine novelty shall be possible. And because they *wish* for such a world they attempt to prove that the world really *is* of this sort. I do not want to be considered here as making a cheap accusation of wishful thinking. Their objections may, for all I know, be philosophically justifiable. I merely wish that this background of emotional revulsion be noted.

Thirdly, it is not altogether clear to me what these philosophers hope for from novelty. Evidently they consider that a universe with novelty in it is very desirable, that it is a fine and grand idea. And evidently they consider that a universe without novelty in it is

NOVELTY, INDETERMINISM, AND EMERGENCE 297

somehow a very poor affair. But why this is so is not self-evident. Novelty *per se* is not a good. That something is new does not seem to be *in itself* a rational recommendation of it. It may be something better or it may be something worse than what we already have. Of course one can understand that the possibility of novelty is also the possibility of improvement, and may therefore give hope. But it is also the possibility of greater anguish and darkness. And I can see no ground, inherent in the bare idea of novelty itself, for betting on the optimistic alternative. It is rational to wish for the better. But is it rational to wish for the new as such? I cannot rid my mind of the impression that these philosophers vaguely and absurdly suppose that novelty is something *per se* desirable. Is it possible that they are influenced by that thirst for change for its own sake, for excitement, for thrills and surprises, which is apt to consume men who have no serious purpose in their lives, men who in consequence suffer from *ennui?* To the man who is bored, any change appears a blessing. Why do these philosophers throw a halo round the concept of novelty? I ask the question and I leave it unanswered.

These remarks are purely desultory. It is not my purpose to *evaluate* the concept of novelty. I shall not again raise the question whether it is rational to desire a world with novelty in it. I will assume, if you like, that it *is* rational, that novelty is a fine thing in itself. Henceforth I shall confine myself to analysis. But I could not refrain altogether from these human reflections. Now that I have got them off my chest I can proceed with my real task.

I do not know who in our own time first emphasized the notion that novelty is a fundamental feature of the world-process, that it is one of the universal categories by which we are to understand our world. Perhaps it was Bergson. Emphasis upon it, at any rate, is comparatively modern; nor is this emphasis found widely in philosophical literature until about thirty or forty years ago.

Now a new concept cannot be left simply suspended in mid air. It has to be linked up in some definite relations with all the other ideas which are already parts of our philosophies. It is here as it is with the process of introducing a new member into a club or a society of people. We have to see whether they harmonize. And much may depend upon the method of introduction. Now what the

28

philosophers of novelty have done, by way of introducing their new idea into the circle of what we may call respectable philosophical society, has been to seek to link it with some other philosophical idea which is already well known, and to get this other idea to introduce it into philosophical respectability. To perform this office three ideas have been successively chosen. These have been the concepts, respectively, of *life*, of *indeterminism*, and of *emergence*. The concept of novelty, being unable by itself to gain an entrance into philosophy, has sought to ally itself with one or other of these and to gain admittance through these alliances. I wish to study each of these three alliances in turn.

It was, indeed, a fairly obvious proceeding to link the notion of novelty with the notion of life. For what the philosophers of novelty were seeking was a means of escape from the dull routine of a mechanical universe. Now mechanism and organism are opposites. They are the opposites of death and life. A mechanical universe, it seems, rules out the possibility of genuine novelty. A living universe, perhaps, would admit its introduction. Life, perhaps, not being mechanical, might well produce, and keep on producing, *new* things. Accordingly, life is declared by Bergson to be the moving force of the cosmic process. This will ensure a universe with novelty in it. The creation of novelties seems to be a characteristic of life. If anyone doubts this let him be reminded of the amazing monsters which now, and in past time, have walked upon the earth or swum in the sea, and of the frightful freaks and abortions which are, from time to time, born into the world. Life is the very creator of novelties. Let our philosophy, then, declare that the universe is actuated by the force of life. We shall then have a universe provided with a constant stream of novelties upon which to feed our emotions of surprise and wonder. This is the motivation of the doctrine of the *élan vital*.

But the attempt to introduce the concept of novelty into philosophy by linking it with the concept of life cannot, without further help from the outside, succeed. By itself this alliance is insufficient to overthrow the philosophy of mechanism. Life, replies the mechanist, only *seems* to produce novelties, and to do unexpected and unpredictable things. It does not really do so. Monsters and freaks impress us, but there is nothing more really new in them than in the strange shapes which mountains, clouds, or other inorganic products sometimes assume. The products of life seem unpredictable because the conditions which govern them are so complicated and subtle that prediction is, in our present state of knowledge, impracticable. The laws of eclipses of the sun and moon are now known. Before they were known an eclipse was unpredictable. The laws of life will some day be discovered. And then the phenomena of life will be predictable. Or at any rate there must *be* such laws, whether anyone ever discovers them or not. We must believe that whatever happens in the universe happens in accordance with law. And if life is governed by law at all, whether known or unknown, then it is reducible to regular causal sequences, and therefore to routine. So once again we shall have a universe in which there will be no novelty in the sense desired by Bergson and James. It should be noted that this argument is not the same as that which is commonly attributed to mechanistic biologists. It does not assert that the laws of life must in the end be reducible to the laws of chemistry or mechanics. That question is quite irrelevant to what we are discussing. The point is that if life is governed by any laws at all, even by special laws of its own which are admitted to be irreducible to lower laws, you will still have a deterministic routine which will exclude real novelty. If life is governed by law, then it is an affair of repeatable sequences, regularities, routine. Hence, even if life is the moving force of the whole cosmic process, yet if the operations of life are governed by law, then everything which happens in the universe will be theoretically predictable, and there will be no genuine novelty in the universe. Hence merely to ally the concept of novelty to that of life does not avail to make of novelty a philosophically tenable conception.

For these reasons the philosophers of novelty have sought to find for their conception a second alliance, this time with the concept of *indeterminism*. Not that novelty will, in order to link itself with indeterminism, give up its alliance with life. Rather it will seek to retain both associations. In other words it will reject the argument that life, being governed by law, can yield no true novelty. Life, it will now be asserted, is not governed by law at all. Life is indeterministic. And therefore it may or will produce novelty. Accordingly both James and Bergson tend to deny that the world is completely deterministic. Both introduce into their philosophies the concept of indeterminism.

The next question we have to face, therefore, is this. Suppose we admit the validity of the concept of indeterminism. Suppose we admit that life is indeterministic. Suppose that we go even further, and admit that the universe at large—and not merely that part of it which is living—is in some radical and fundamental way indeterministic, will this really validate the concept of novelty? In fine, if the world is indeterministic, does this imply that the world will contain novelties?

We cannot proceed any further with our enquiries until we have attempted to give an answer to a question which, I hope, has been puzzling my readers for some time. This is the question, what is *meant* by the concept "novelty"? What is the proper analysis of the concept? To give a complete and final analysis of it is a task which I shall not attempt to carry out. For I think that the matter can be sufficiently clarified for our purposes in this paper if I make a distinction between two possible senses of the word novelty. When we speak of novelty, I will say, we may be thinking either of *absolute* novelty, or of *relative* novelty. Suppose we have a causal sequence A . . . B. Suppose the cause A is an explosion, and the effect B is my death. Now in one possible sense of the word there is present here the arising of novelty. For death is very unlike an explosion. The effect B is something which is different from, and very unlike, the cause A. Something *new* has therefore arisen, something, that is, which was not there before. In fact there must be something new coming into being in every case of change. Change implies that something comes into being which was not in being before. If what is red turns green, then the green is something new. It is a novelty. But this kind of novelty is merely relative. The new elements of such a situation are new in that situation and relatively to that situation. But they are not absolutely new entities in the universe, for they may have appeared in the universe millions of times before. This is the case with death and with the appearance of the color green. I would only call anything an absolute novelty if it were a phenomenon the like of which had never appeared before in the whole history of the universe. Suppose that up to the present time there never had existed anything green in the universe, and that now suddenly something green should appear. Then I should call green color an absolute novelty. Or suppose that there was a time, many millions of years ago, when life first made its appearance in the world. Then life would have been at that time an absolute novelty in the universe.

Now which kind of novelty is it that James and Bergson would have us believe in? I think it is quite obvious that what they are thinking of is what I have called absolute novelty. There would have been no point whatever in insisting that the world should contain relative novelty. For everyone who admits the existence of change would have to admit the existence of relative novelty. Only philosophers who deny the existence of change, the Eleatics for example, could possibly deny that relative novelty occurs in the world. Obviously what James and Bergson wanted was a universe which might from time to time throw up existences and experiences which should be utterly new. And it is evident that this is also what is meant by those emergent evolutionists who claim that emergence ensures novelty. In the system of Alexander, for example, the primitive stuff, space-time, suddenly gives rise to the secondary qualities of matter. Later on life emerges, and later still mind. These new qualities, or existents, are declared by Alexander to be novel precisely in the sense that they had never, until the moment of their emergence, appeared in the universe before. Clearly then we are not concerned with mere relative novelty. And we may define the concept of novelty which we are discussing in this paper to mean the arising of something hitherto unknown in the history of the universe. I admit that this a very vague definition, and that all sorts of questions might be raised about it. But it is, I think, sufficiently clear for our purposes here.

The question which we have to ask therefore is: Does indeterminism imply, or even make possible, novelty in this sense of absolute novelty? Clearly James and Bergson thought it does. Now I take it that the notion of indeterminism might be explained in some such way as this. In a deterministic world every event will be completely determined by its causes, so that if A . . . B constitutes a causal sequence, then whenever A happens B will happen, B and nothing else. But if there were a world in which A might sometimes be followed by B, sometimes by C, sometimes by

D, and so on, and in which absolutely no antecedent conditions existed to determine *which* of these events should follow A, such a world would be to that extent indeterministic. It is alleged, whether truly or not I do not know, that the sub-atomic world is, in certain respects, like this. Given a certain set of circumstances X, then the electron may jump either to the right or to the left, and there is nothing in the antecedent set of circumstances X to determine which way it will jump. Let us assume for the sake of example that the world is, either in regard to the whole of it or in regard to some part of it, indeterministic, in this sort of sense. Then the question is, Will such indeterminism carry with it novelty of the kind demanded by James and Bergson?

Let us revert to the example of the electron. Even if *all* the relevant antecedent circumstances are completely known, we are to assume, it is still uncertain whether the electron will jump to the left or to the right. It may be that for some reason or other I *expect* it to jump to the left. Actually it jumps to the right. Now where, I ask you, is the novelty? In what way is a jump to the right more intrinsically novel than a jump to the left? Has there been introduced into the world anything whatever, entity, quality, or experience, which has never occurred in the universe before? Obviously not. Objects moving right, left, and in all other directions are perfectly familiar features of experience. There is nothing new in an object moving to the right. Of course, if we expected it to move to the left, then what has happened is something *unexpected*. But the unexpected is not the same as the new. An indeterministic world would no doubt be a world of constant surprises. But there is no guarantee that anything *new* would ever arise in it. It might be the case, even in a completely indeterministic world, that nothing ever occurred except the same old experiences repeated *ad infinitum,* although these old experiences might keep turning up in the most unexpected places and times. No one would ever know what was going to happen next. But this would not imply that what was going to happen next would be something which never happened before. To argue in this way would be a complete *non-sequitur*. From the fact that I do not know what is going to happen tomorrow you cannot argue that something novel will occur. Clearly there is no necessary connection whatever between the conception of indeterminism and the conception of novelty.

The philosopher of novelty may, however—while admitting that indeterminism does not logically or necessarily *imply* novelty— suggest that at least it renders novelty *possible*. In a deterministic world, he may say, novelty is downright impossible. We know beforehand that it cannot occur. But in an indeterministic world, though we cannot be certain that there will be novelty, yet it is at least possible that there might be. And therefore it is desirable to believe in indeterminism if we want to believe in novelty. Determinism rules out the possibility of novelty, while indeterminism does not.

Unfortunately, however, this contrast between determinism as rendering novelty impossible and indeterminism as rendering it possible is quite false. For in point of fact novelty is just as much possible in a deterministic world as in an indeterministic one. This may be seen in two ways. First, suppose that in a completely deterministic world you have the following case. There exists in the sun a chemical element X, which does not exist anywhere else in the physical universe. And there exists in the star Canopus another chemical element Y which also is not duplicated anywhere else in the world. Owing to their spatial separation these two elements X and Y may have never in the whole history of the world come into chemical combination with one another. Now suppose that tomorrow they are brought together. They combine and form a compound substance which has never before existed in history. This will be a case of absolute novelty, and it will have occurred in a physical world governed by known chemical laws, in a universe wholly deterministic. Indeed without going so far afield as Canopus and the sun I see no reason to think it impossible that our earthly chemists do, or at least might, occasionally bring together and combine elements found on the earth which have never been combined in nature. In that case too we shall have absolute novelty combined with determinism.

There is a second way in which the same point may be made. We find in such a system as that of Alexander an actual example of a philosophy which successfully and quite self-consistently combines determinism and novelty in the same world-picture. Alexander's universe is completely deterministic. And yet according to him novelty appears every time that what he calls a new "quality" *emerges* in the universe. For example, when the motions within space-time become sufficiently complex, the secondary qualities emerge and with them matter comes into existence. The emergence of matter was completely determined by the previous motions in space-time. And yet it was an absolute novelty in the universe. Again when the motions within the living organism become sufficiently complex, the quality of consciousness emerges. The sudden appearance of mind in the world is something utterly new, and yet it was completely determined by the previous motions in the nervous system of the organism. We see from these examples that novelty is quite compatible with determinism. And therefore it is false that novelty is any more possible in an indeterministic world than it is in a deterministic one.

The conclusion which I reach is that there is absolutely no connection whatever between the two concepts of novelty and indeterminism. Indeterminism does not imply novelty. Novelty does not imply indeterminism, since it may occur in a deterministic world. Indeterminism does nothing to render novelty either possible or probable. Thus the two concepts are entirely independent. The supposition of James and Bergson that they are connected is a sheer delusion. They are in no sense friends or natural allies. Hence the concept of novelty cannot be smuggled into polite philosophical society by pretending that it is a friend or relative of indeterminism.

What is the source of this curious delusion on the part of James and Bergson? The explanation, I think, is as follows. What is introduced into the world by indeterminism is not novelty but— what is quite a different thing—*unexpectedness*. The difference between the two is this. Novelty, if there is such a thing, is an objective character of the world. If anything occurs which has never occurred before, then it is an objective fact that it never has occurred before. But unexpectedness is merely a subjective attitude of an observer's mind. Now those who praise indeterminism in the belief that it will yield novelty are simply confusing unexpectedness with novelty. It is quite true that, in an indeterministic world, one would not know what to expect, and therefore occurrences would be unexpected. And the philosophers of novelty have falsely concluded that the occurrences would therefore be novel. They have projected their subjective state of surprise at the unexpected upon the objective world and called it novelty. And the cause of this confusion seems to me to be one of the most elementary fallacies in logic, that, namely, of simply converting a universal affirmative proposition. All novel things are surprising. Therefore, it has been thought, all surprising things must be novel.

The attempts to introduce the concept of novelty into philosophy by allying it first with the concept of life and then with that of indeterminism have thus proved to be failures. I pass now to the third concept with which it has been sought to ally it, namely that of emergence. And I shall take as my example of this the philosophy of Alexander.

It will be remembered that, according to Alexander, the world is entirely deterministic. Everything that happens is determined by its antecedent conditions. But there are two possible cases. First, a set of conditions X and their resultant Y may both be on the same level of existence. In that case the resultant is predictable in terms of its conditions, and we have no novelty. Thus billiard balls impacting cause certain changes of motion. But the effect is entirely predictable and nothing radically new comes into existence. The cause is mechanical motion and the effect is mechanical motion. But secondly the set of conditions X may be on one emergent level, the resultant Y may be on a higher level. If so we have a case of emergence, and also a case of novelty. Thus a set of motions in empty space-time has as its resultant the sudden appearance of matter in the world. Matter has emerged, and matter is at the same time something completely new in the universe. Again, a set of motions in a living organism has as its resultant the emergence of consciousness which, when it first appears, is an absolute novelty. In this way the concept of novelty is linked with the concept of emergence. Emergence produces novelty.

Now the idea of emergence is, as someone has pointed out, a development of Mill's distinction between homopathic and heteropathic effects. A homopathic effect is one which is compounded out of its causes, and is predictable in terms of them. A heteropathic effect is one which is not. For example, if a billiard ball strikes another similar ball, at a certain angle and with a given velocity, the resulting motions, directions, and velocities can be predicted. This is a homopathic effect. But if two colorless liquids are mixed in the laboratory, the mixture may turn suddenly deep blue. A blue liquid could not have been predicted as a resultant of two white liquids. This is therefore a heteropathic effect. The concept of the heteropathic effect is practically identical with the concept of emergence. In the impact of the billiard balls we have cause and effect on the same level, we have mere rearrangement of old elements in new patterns, we have predictability. In the blue liquid we have something new, something which could not be found in the two white liquids, something therefore which is unpredictable.

But this distinction between heteropathic or emergent effects and homopathic or non-emergent effects is not tenable. Emergentists make the distinction twofold. They say (1) that non-emergent effects are predictable, emergent effects unpredictable. And they say (2) that emergent effects are novel, non-emergent effects non-novel. Neither of these contrasts can be maintained. First, as to predictability. The heteropathic effect, the blue liquid, it is said, could not be predicted from the two white liquids without experience. Once I have experienced the sequence I can, of course, predict its recurrence, relying on the uniformity of nature. But *without experience* I could never predict it. This is certainly quite true. But is it not equally true that, without experience of impacts, I could not predict the angles or velocities of the new motions of the billiard balls? Indeed I could not even predict that there would be any motions at all. When a moving billiard ball strikes a ball at rest, why should both not thereupon stand still? Or why should they not turn and go backwards upon their tracks? For the matter of that why should they not turn into watermelons, or disappear out of existence altogether? One can give no reason why they do what they do, or why they do not do any of these other strange things. One has to wait on experience to find out what will happen as the result of an impact just as much as to find out what will happen when one mixes two white liquids. Once we have experienced the sequence, in either case, then of course we can predict future sequences of the same kind upon the basis of the uniformity of nature.

I am, of course, merely repeating here considerations which have been familiar to everyone since the time of Hume. And it is accordingly unnecessary for me to elaborate them. It is only necessary to note that what Hume said really disposes of the alleged difference between emergent and non-emergent effects in the point of predictability. Both are equally predictable after experience. Both are equally unpredictable before experience.

I now turn to the second point of alleged distinction between them. Emergent effects are said to possess a novelty which is absent from non-emergent effects. This distinction is also untenable. The two white liquids become blue when mixed. The blue was not there before, it is said, and therefore this is something new. There is no novelty of this kind in the case of the impacting billiard balls. The cause in that case is motion, and the effect is also motion. Cause and effect are essentially alike. Nothing new enters upon the scene.

But I cannot admit this contrast. In all cases of cause and effect there are points of difference between the cause and the effect, and also there are points of resemblance. The supposed distinction between the two kinds of effect which we are discussing arises from arbitrarily ignoring the likenesses in one case and the unlikenesses in the other. In the case of the billiard balls we concentrate on the likenesses between cause and effect, ignoring the differences, and then declare that the effect entirely resembles the cause. In the case of the liquids we concentrate on a single point of unlikeness, that of color; we ignore the points of resemblance, and then declare that the effect is unlike the cause and that we have here a case of novelty.

For consider. What is it that is unlike in the case of the liquids? Nothing except the color. All the other qualities in the cause and effect are the same. The cause, the unmixed liquids, has the quality of liquidity. So has the effect. The unmixed liquids have a certain volume and weight. The volume and weight of the mixture is merely the combined volumes and weights of its components. Nothing is changed except the color. Only one quality is different as between cause and effect, all the rest remaining the same.

Now take the case of the billiard balls. It is true that there is a resemblance between cause and effect which is expressed by saying that both are motion. This corresponds to the resemblance in the other example which was expressed by saying that both cause and effects are liquids. But the *velocities* of the two balls will be changed by the impact, and so also (in the majority of cases) will the *direction* of motion. In the case of the liquids we have a change of color. In the case of the billiard balls we have a change of velocities and direction.

Now why is a change of color a novelty, and a change of velocity and direction not a novelty? Would it be rational to say that a change of color involves novelty, but that a change of smell does not? And can there be any more justification for saying that a new color is novel, but that a new velocity and direction are not? I cannot see the slightest justification for such a distinction in any of these cases. If I am right, this second distinction between emergent and non-emergent effects collapses as certainly as did the first.

Why is it, then, that we all tend to think that there *is* some distinction? For I think it does seem to most of us that the case of the liquids is somehow unlike the case of the billiard balls. This is, I believe, a sheer illusion, and I account for it as follows. I suggest that what are called homopathic or non-emergent effects are simply those which are more common in our experience, more familiar, less surprising, less striking. The so-called heteropathic, or emergent, effects are those which are comparatively rare, so that when they do occur they seem more striking and unexpected, and so we get the impression that we are in the presence of some kind of novelty which is absent in the other cases. Every human being, even every savage, has experience of bodies impacting. The effects, therefore, are so familiar and expected that we think we could predict them and that there is no novelty in them. But the experience of the two white liquids becoming blue is a very rare experience in comparison with the other. In nature it is scarcely ever seen, but only in the laboratory. The savage has probably never seen it or anything like it. Even to the civilized man, unless he happens to be a chemist, it is a very odd and unusual experience. Most of us have seen it perhaps once or twice in our lives when we were schoolboys and spent our weekly hour of relaxation and amusement amid explosions and magical transformations in the 'stinks' lab. Or perhaps we have only seen it when, as adults, we have been shown, as gaping and astonished laymen, through a chemical laboratory. This effect, then, is so unfamiliar, so striking, that it arrests our attention. We fasten on this difference between cause and effect and forget the similarities. This change of color has the quality of being surprising, unexpected. And as happened in the case of indeterminism, we confuse the unexpected with the novel. We think that this possesses a novelty which is lacking in the case of the impacting billiard balls. In short, the difference between the two kinds of causation, the emergent and the non-emergent, is not objective at all. It is subjective. It is a difference in our mental attitudes.

I do not deny that absolute novelties may appear in the world. Life and mind were perhaps once such. If the concept of emergence is merely intended to draw attention to this fact, it is unobjectionable. But it is false in so far as it implies a radical distinction between emergent effects as novel and unpredictable and non-emergent effects as predictable and non-novel. Absolute novelty will arise whenever an event or entity appears which has never appeared before. And this may happen either as a so-called emergent effect or as a so-called non-emergent effect. On the first occasion on which life or mind appeared we should have absolute novelty. And this would ordinarily be classed as an emergent. But it is also true that on the first occasion on which a new velocity or direction of motion occurred we also have absolute novelty, although this would ordinarily be classed as a non-emergent effect. The only difference between these cases is that, for subjective reasons, the sudden appearance of life or mind in the world seems to us very striking and important, while the sudden appearance of a new velocity seems to us trivial and unimportant. Thus emergence seems to dwindle into unimportance as a philosophical concept. At any rate it ought to be clear by now—and this is my main point—that the concept of novelty has nothing to hope for from an alliance with the concept of emergence. We can admit the possible appearance of absolute novelties, such as life and mind, in the world without any theory of emergence at all. Therefore the theory of emergence does not really help the cause of those who are anxious for a universe in which novelty is possible.

Nothing in this paper is to be construed as an attempt to show

40

that absolute novelty is impossible. On the contrary the result is the very opposite. The philosophers of novelty thought that they could not have novelty in the world unless they could prove either that life is the principle of things, or that the world is indeterministic, or that there is emergence. Our analysis has shown that none of these alliances and connections is essential to the idea of novelty, that it is dependent on none of these conditions. It can stand on its own feet. There seems no reason to doubt that absolute novelties do come into existence. Life and mind must, I suppose, have come into existence at some time. And even in the chemical laboratory I dare say it may be the case that new chemical compounds with new qualities sometimes make their appearance. And what I have attempted to show is that absolute novelty may exist whether the world is living or dead, whether it is deterministic or indeterministic, whether it is mechanical or organic, whether there is or is not such a thing as emergence.

I said that I would not attempt to evaluate the concept of novelty. But I cannot refrain from pointing out that if our analysis is correct, it loses most of the significance which its authors attached to it. Just because it is consistent with almost any philosophy, with idealism or materialism, with an organic philosophy or a mechanical philosophy, with determinism or indeterminism, with a living world or a dead world, it is reduced, in my opinion, to a very low degree of importance. The philosophers of novelty thought that by means of their new conception they could relieve us of the crushing weight of a dead soulless mechanical universe. They could introduce us to a world instinct with poetry and life. This, I think, was their main object. But if I am right, they have failed in this aim. For even if there be novelty in the world, in spite of it the world may yet be nothing but a soulless machine. Of course I am not saying that it is so. That is not the point. For all I know the world may be alive, organic, creative, soulful, poetical, noble. But the concept of novelty does not help us to show that it is so.

W. T. Stace

Princeton University

3. The philosophic functions of emergence
Charles A. Baylis

Originally published as Baylis, C. A. (1929). "The philosophic functions of emergence," *The Philosophical Review*, ISSN 0031-8108, 38(4): 372-384. The *E:CO* editorial team would like to thank Duke University Press, especially Thomas Robinson, for their kind permission to reprint this article.

Emergence is not ordinary change: Introduction to Baylis

The idea of emergence in its complexity science sense was first broached by adherents of Emergent Evolutionism, a loosely joined movement of scientists, philosophers, historians, social thinkers, and even theologians during the first quarter of the twentieth century (Blitz, 1992). Included among the proponents of Emergent Evolutionism in England were the animal behaviorist C. Lloyd Morgan and the philosophers Samuel Alexander, C. D. Broad, and Alfred North Whitehead (the eminent mathematician had turned to philosophy in his later years). In the US, Emergent Evolutionism was promulgated by the entomologist W. H. Wheeler (who incidentally was one of E. O. Wilson's teachers), the philosophers Roy Wood Sellars and John Boodin, the philosopher and historian Arthur Lovejoy, and the social theorist and philosopher George Herbert Mead (quite influential to the esteemed contemporary philosopher Jürgen Habermas).

Baylis's four types of emergent change

Not surprisingly, criticisms of the concept of emergence began right on the heel of this movement. A common objection was voiced against the claim of emergence being impervious to reductionist or mechanistic explanations. One strategy of the detractors along this line was to take on those purported characteristics of emergent phenomena that enabled it to be protected from reductionist onslaught, e.g., the property of radical novelty of emergent level phenomena in relation to the lower, substrate level (Goldstein, 1999). Such was the tack taken in 1929 by the American philosopher Charles Baylis in the following article. In particular, Baylis, who taught at Brown and the University of Maryland, believed that he had, by puncturing the claim of radical novelty, demonstrated that emergence was no different than the type of discontinuity accompanying ordinary change and therefore could be explained by causal mechanisms that were typically applied to ordinary change.

To advance his argument, Baylis argued that there were only four possibilities of emergent change:

1. *Integrative emergence:* when a new property appears as the result of an integration of components, e.g., at room temperature, hydrogen and oxygen are gases yet when chemically combined the result is the new property of liquidity;

2. *Integrative submergence:* when properties of the parts are lost in the resulting whole, e.g., water (again) in which the gaseous property of the component elements is discontinued in the liquidity of the whole;

3. *Disintegrative emergence:* when new properties of the parts emerge when they are separated from the whole of which they were previously a part, e.g., what happens when oxygen and hydrogen are separated from water;

4. *Disintegrative subemergence* when the properties the parts had when combined are lost when they become separated from the whole, e.g, when hydrogen and oxygen are separated from water and thus lose property of liquidity. (To be sure, points 3 and 4 appear to be two sides of the same coin).

For example, Baylis referred to the complex $(a+b)$, which, having the character of being greater than both a and b alone, could be considered an instance of 'integrative' emergence since it had a character neither of its elements had alone. Furthermore, if this same complex were dissolved, its unique character as a complex would 'disintegrate' thus losing the previous properties it had as a whole complex. However, and this was a key move on the part of Baylis, from another perspective, the supposedly *integrative* complex could at the same time be seen as *disintegrative* in that the properties that a or b alone might have had before they were combined into the complex $(a+b)$ would be lost when a and b became part of the complex $(a+b)$. Effectively, what Baylis had considered was how an emergent whole could be considered *less than the sum of its parts*!

Since these four types do indeed appear to have encompassed all the possibilities of what can take place during changes from parts to wholes and vice versa, Baylis was arguing that the discontinuous novelty characterizing emergent phenomena was ubiquitous to all changes and therefore it was arbitrary to identify emergence only with the first type, integrative emergence. To drive his point home,

Baylis considered the change resulting from moving a book from one shelf to another. The change of place of the book could be considered either an integrative or a disintegrative novelty depending on your point of view: a new *integration* exhibited in the novel pattern of books at the new location or a new *disintegration* of the pattern on the shelf before the book was moved. Baylis concluded that novelty was thereby constituted by either a building-up of integrations or their tearing down. Consequently, the production of discontinuous novelty by itself didn't aid our understanding of so-called emergent evolutionary processes.

Furthermore, since it was arbitrary as to whether a given change was to be considered an integration or a disintegration, Baylis believed the erection of any particular hierarchy of putatively emergent levels was an entirely discretionary matter, emergentists merely picking out and hierarchically arranging just those levels which interested them according to their own agendas. Because novelty was so pervasive with all change, moreover, a new level could be assigned every time something new came about and thus leading to a pervasive infinity of levels, an idea which Baylis found absurd. Or as the philosopher C. W. Berenda (1953) once blithely put it in commenting on the idea of emergence in the nineteen fifties, "There is nothing new in novelty"!

Stace's causal chain

We can take Baylis's argument, however, in another way, namely as having set the bar higher as to what should count as genuinely emergent novelty and not merely that kind of novelty which accompanies ordinary change. That is, by pointing out the arbitrariness as to what to count as emergent novelty, we can take Baylis as having demonstrated that the early emergentists had not sufficiently qualified emergent novelty. This suggestion will become clearer by looking at a similar argument offered around the same time as Baylis by another critic of proto-emergentism, the philosopher W. T. Stace (1939; incidentally, Stace achieved his fifteen minutes of fame much later for arguing against the religious authenticity of psychedelic drug-induced mystical experience, see Stace, 1960). Similar to Baylis's concern with discontinuity during change, Stace's dispute with emergence focused on the nature of causal chains operative during any process of change. In the causal sequence *A*... *B*, if *B* was discontinuous with *A*, then there was something new by definition in *B*. But, since every time there was a causal sequence, the focus could be either on how the effect was novel *or* how the effect was similar with respect to the cause (since causality implied both a difference and a similarity between cause and effect), Stace held emergentists to the task of justifying why they pushed for emergent phenomena being novel and not the same as antecedent conditions. He concluded that the only reason for emphasizing discontinuity had to do with the relative commonness or rarity of an experience: "...when [emergents] do occur they seem more striking and unexpected, and so we get the impression that we are in the presence of some kind of novelty which is absent in the other cases" (Stace, 1939: 308). Indeed, many unexpected things can happen, but their unexpectedness might merely be old things happening in unexpected places. The sense of unexpectedness was, after all, a subjective attitude on the part of an observer's mind, a sense of surprise projected into a claim of novelty. The sense of novelty, for Stace, therefore said more about the person beholding an emergent than the emergent itself. Moreover, according to Stace, novelty as such was compatible with many philosophical positions, deterministic and indeterministic, mechanism and vitalism: "For all I know the world may be alive, organic, creative, soulful, poetical, noble. But the concept of novelty does not help us to show that it is so" (Stace, 1939: 310). As a result, Stace held that novelty was merely epistemological and had no real bearing in establishing the ontological status of emergence.

The contextuality of emergent novelty

The arguments of both Baylis and Stace can be explicated along the lines of the mathematical construct of equivalence classes in which an operation is devised and a definition of equivalence and inequivalence is posited. The equivalence class is constituted by those members defined to be equivalent for that particular example, but it is an arbitrary designation depending on how one defines the operation and the equivalency. For example, in the operation known as congruency, one number is defined as equivalent to another according to whether it has the same remainder after being divided by some arbitrary number (Ash, 1998): if the operation is defined as division of an integer by 4, then one possible equivalence class would be those multiples of 4 that when divided by 4 leave a remainder of 0, e.g., the class made-up of (...-8, -4, 0, 4, 8 ...). Notice that equivalency in this context does not imply identity which would be the complete absence of novelty. That is, -8 is not identical to 4 but, instead, for the particular operation of dividing by 4 and resulting in 0 as a remainder, -8 and 4 are equivalent, or, what amounts to the same thing, not different with respect to that operation. The point is that what is considered the same, i.e., equivalent, and what is considered not equivalent, i.e., novel, are arbitrary markers depending on how one has first defined what is to count as equivalent or not.

Besides the element of arbitrariness involved in deciding what's novel, there is another aspect of the contextual nature of novelty, namely, that something is either novel or similar with respect to another thing according to how that same thing is novel or similar with respect to another thing altogether. For example, British English and American English are distinct, i.e., novel with respect to each other on many points, yet they are more similar to each other than they both are to French. Similarly, English is more similar to French than French or English are with respect to Japanese. The point again is that the designation of something as novel is contextual and this must be made explicit when specifically considering *emergent* novelty.

Getting back to the point made above about taking Baylis (and now Stace) as having set the bar high for what is to count as genuinely emergent novelty, it must be recognized that a serious question-begging flaw can be detected in arguments that go from the contextuality involved in defining novelty to the dismissal of emergence by being identical with ordinary change. As a matter of fact, both Baylis and Stace started off their discussions of emergence by first identifying it with ordinary change in general: Baylis did this by conflating emergence with his four types of change and Stace by understanding emergence within the framework of change as a causal chain. But once these identifications have been made, then, of course, any novelty characterizing emergent phenomena will essentially be the same as the novelty accompanying change in general.

In any effect, we now have a bar over which emergence must be defined: emergence is not ordinary change in general but is instead consonant with a special kind of change, i.e., one that generates the outcomes which are unpredictable, non-deducible, irreducible, and capable of daunting (not violating) traditional notions of causality and determinism (Kekes, 1966). That is, if the construct of emergence is to indeed have the significance emergentists want it to have, it must be referring to a very different kind of change than the ordinary kind. As the philosopher Paul Henle (1942) pointed out over fifty years ago, the novelty found in doctrines of emergence amounted to much more than, say, a new automobile coming out of an assembly line in that not just the actual matter making up a car itself must be new, the *form* must also be novel.

A failure of the imagination

The criticisms of emergence leveled by Baylis and Stace both can be said to have rested on a similar picture of the process of natural change. We can see Baylis's picture in what he appealed to in order to explain how novelty could be associated with every change: the process of moving a book from one shelf to another. To be sure, such a change does fit with Baylis's definition of the novelty of emergent 'complexes' which he equated with the 'gestalts' of Gestalt psychology. The process of moving a book from one shelf to another leads to not only a new pattern on the new shelf, but a new pattern on the old shelf. But there is nothing unpredictable or irreducible about these two instances of novelty. The book is easily taken back from its new place to the old one, thereby, obliterating the novelty the original movement brought about. By understanding emergence in the specific terms of how he conceived change and novelty, Baylis could maintain that there was both a type of symmetry and arbitrariness between integration and disintegration. Moreover, Baylis's process of emergence thereby shared the attribute of reversibility. Similarly, Stace's simple picture of the causal chain had no place in it for the nonlinear, complexifying operations going on in complex systems. In both cases we see a failure of the imagination as to what is possible for nature. In the age of complexity, such a failure of the imagination can no longer get by without notice.

Jeffrey A. Goldstein

References

Ash, R. (1998). *A primer of abstract mathematics*, Washington, DC: The Mathematical Association of America, ISBN 0883857081.

Berenda, C. W. (1953). "On emergence and prediction," *Journal of philosophy,* ISSN 0022-362X, 50: 269-74.

Blitz, D. (1992). *Emergent evolution: Qualitative novelty and the levels of reality*, Dordrecht, The Netherlands: Kluwer Academic Publishers, ISBN 0792316584.

Goldstein, J. (1999). "Emergence as a construct: History and issues," *Emergence*, ISSN 1521-3250, 1(1): 49-62.

Henle, P. (1942). "The status of emergence," *Journal of philosophy,* ISSN 0022-362X, 39(18): 486-493.

Kekes, (1966). "Physicalism, the identity theory and the doctrine of emergence," *Philosophy of science*, ISSN 0031-8248, 33: 360-75.

Stace, W. T. (1939). "Novelty, indeterminism, and emergence," *Philosophical review*, ISSN 0031-8108, 48: 296-310.

Stace, W. T. (1960). *Mysticism and philosophy*, New York, NY: MacMillan, ISBN 0874774160 (1987).

THE PHILOSOPHIC FUNCTIONS OF EMERGENCE.

1. THE concept of emergence has been given a position of great importance by noted contemporary philosophers. Under one form or another and with varying names it has been made an integral part of the metaphysical views of such diverse thinkers as Bergson, S. Alexander, Lloyd Morgan, Broad, Baldwin, Hobhouse, Laird, Sellars, Spaulding, H. C. Brown, and Lovejoy. Some of these have linked it with the concept of evolution and made it the basis of their whole system, for example, Lloyd Morgan in *Emergent Evolution,* R. W. Sellars in *Evolutionary Naturalism,* and S. Alexander in *Space, Time, and Deity.* Others have thought it offered solutions to particular philosophical problems. Lovejoy [1] and Alexander [2] use it to solve the problem of the status of secondary qualities, though in different ways. Lloyd Morgan [3] thinks it clearly refutes the contention of the New Realists that nothing accrues to an object when it is perceived. Others have thought that emergence has solved the mind-body problem; witness Broad's " Emergent Materialism ",[4] H. C. Brown, " The Material World—Snark or Boojum ",[5] and Sellars, *Evolutionary Naturalism,* not to mention Lloyd Morgan's own book. Similarly the concept of emergence has been used to resolve the mechanism-vitalism controversy and a number of other metaphysical problems.

The aim of this paper is to point out that, in spite of the fact that emergence is more widespread than even its most ardent advocates have claimed, for it is indeed ubiquitous, nevertheless it solves none of these problems, supports no one *Weltanschauung* rather than any other, and does not even imply evolution.

[1] A. O. Lovejoy, *The Discontinuities of Evolution,* in *Essays in Metaphysics,* University of California Publications in Philosophy, Vol. 5, 1924, pp. 180, 187–188.

[2] S. Alexander, *Space, Time, and Deity,* Vol. II, p. 59.

[3] C. Lloyd Morgan, *Emergent Evolution,* p. 99.

[4] C. D. Broad, *The Mind and Its Place in Nature,* p. 646.

[5] *The Journal of Philosophy,* Vol. XXII, No. 8, April 9, 1925, pp. 197–214.

372

PHILOSOPHIC FUNCTIONS OF EMERGENCE.   373

2. Although first made popular by Lloyd Morgan with his happy name for it, " Emergence ", the concept thus signified has been made use of under various names at least since the time of G. H. Lewes.[6]   Many different definitions of it have been offered, but these reduce to two main types, the one, common in England, exemplified by Broad, the other, common in America, exemplified by E. B. Spaulding.

Broad writes, " Put in abstract terms the emergent theory asserts that there are certain wholes, composed (say) of constituents A, B, and C in a relation R to each other; that all wholes composed of constituents of the same kind as A, B, and C in relations of the same kind as R have certain characteristic properties; that A, B, and C are capable of occurring in other kinds of complex wholes where the relation is not of the same kind as R; and that the characteristic properties of the whole $R(A, B, C)$ cannot, even in theory, be deduced from the most complete knowledge of the properties of A, B, and C in isolation or in other wholes which are not of the form $R(A, B, C)$."[7]

In America the term ' creative synthesis ', at least until recently, has been more commonly used, and creative synthetic wholes have been defined as wholes which are more than the sum of their parts. Or, as Spaulding puts it, " In the physical world (and elsewhere) it is an established empirical fact that parts as *non-additively* organized form a *whole* which has characteristics that are *qualitatively different* from the characteristics of the parts. . . . This process of the formation of new qualities through the organization of parts into wholes may be called creative synthesis."[8]

A more precise definition, which includes both of these and yet has certain advantages over them, is here proposed: *Emergents are those characters of a complex which are not also characters of a proper part of that complex, and emergence or creative synthesis is that event which occurs when a complex having emergent characters is formed.*[9]   To take a concrete example, according to this

[6] G. H. Lewes, *Problems of Life and Mind*, First Series, Vol. II, Prob. V, ch. iii, pp. 412–414.

[7] C. D. Broad, *op. cit.*, p. 61.

[8] E. G. Spaulding, *The New Rationalism*, pp. 447–448.

[9] In this definition the term ' complex ' (made up of ' elements ') is used as a general term meant to include all such entities as ' wholes ' (made up of

definition the synthesis of hydrogen and oxygen in the proportions H_2O is a creative synthesis because this compound has characters which neither hydrogen nor oxygen nor the relation of chemical combination has, such as liquidity, its specific heat, and its density.

This definition gives, on the face of it, the essence of the American way of speaking of the concept, but in much more precise terms. Furthermore, it includes the requirement of Broad's definition without introducing the notion of deducibility, or, as Broad more frequently calls it, predictability. For, in the case of empirically knowable complexes (and that is what Broad is talking about), it is clear that characters of the complex which are not characters of its elements could not have been deduced from or predicted from knowledge of the elements alone but require at least one case of knowledge of the complex.

The definition of this paper, in addition to its preciseness and its inclusiveness of current formulations of the meaning of emergence, has the advantage of making the concept applicable to logical and mathematical as well as to physical and mental entities. For example, the logical complex, " p materially implies q ", has the emergent character of being true when both p is false and q is false, for none of the elements of this complex have this character.

3. Much attention has been called to the *emergence* of new characters upon the formation of a complex, but the fact has been neglected that such an occurrence is accompanied by the *submergence* of some of the characters of the elements that formed the complex. To take a concrete case, when hydrogen and oxygen combine according to the formula H_2O it is true not only that this compound has emergent characters but also that the uncompounded hydrogen and oxygen lose, in the process of combining, some of the characters which they possessed uncompounded, such as the character of being a gas. If the new characters which the complex exhibits are called emergents, the old characters which the elements lose may properly be called submergents and the process in which they are lost submergence.

Another fact that has been neglected is this: Emergence and

'parts'), 'compounds' (made up of 'constituents'), 'classes' (made up of 'members'), etc.; and the term 'character' is used as a general term meant to include all such entities as 'qualities', 'properties', and 'relations'.

submergence result from disintegration as well as from integration. For example, disintegrate water into hydrogen and oxygen. The disjoined hydrogen and oxygen now exhibit characters which they did not possess as members of the compound, for example, the character of being a gas. These emerge, whereas some of the characters of the compound (those which were in the reverse process called emergents), liquidity, for example, are submerged. The most convenient terms by which to denote the four possible processes would seem to be *integrational emergence, disintegrational emergence, integrational submergence,* and *disintegrational submergence.*

It might be urged that only integrational emergents and not disintegrational are entitled to the name emergents, because, at least in the case of physical complexes, those characters which a complex but none of its elements has, are new not only in the sense that the elements have no such characters and the complex does, but also in the sense that their occurrence as characters of the complex could not have been predicted merely from knowledge of the elements but required also knowledge of at least one case of the complex. But this is equally true of disintegrational emergents in the physical realm. Those physical characters which a complex does not possess, but which its elements when not integrated in this complex do possess, are not predictable from knowledge of the complex alone but require also knowledge of at least one case of the non-integrated elements. If the emergent characters of water cannot be predicted merely from knowledge of uncompounded hydrogen and oxygen, but require knowledge of at least one case of water, similarly the emergent characters of hydrogen and oxygen cannot be predicted merely from knowledge of water but require knowledge of at least one case of hydrogen and oxygen uncompounded. The cases are perfectly parallel.

4. The example of H_2O shows that integrational and disintegrational emergence and submergence do occur, at least in the physical realm. They occur also in the mathematical or logical realm, as an example will indicate. Consider the complex $(a + b)$. It has the emergent character of being greater than a and greater than b, for this is a character which none of its elements have (integrational emergence). If the complex is dissolved this character is

submerged (disintegrational submergence). On the other hand, when the complex was formed each of the elements lost, *qua* member of the complex, the character it previously had of being, say, multipliable by two independently of the other (integrational submergence). When the complex is broken up they each regain this quality, that is, it emerges again (disintegrational emergence).

Emergence and submergence, integrational and disintegrational, occur in the realms of physical, mental and logical entities, not merely occasionally but ubiquitously. They accompany every change in the universe.

In the first place, since by the term 'complex' is meant any n entities in any n-adic relation, it follows that every change in the universe creates some complexes and destroys others, *i.e.,* integrates some elements into a complex and disintegrates some complexes into their elements, for each such change means the breaking of old relations and the forming of new. For example, I move Kant's *Critique* from the second to the third shelf of my bookcase. The complex constituted by the color pattern of the books formerly on the second shelf is destroyed, and the complex constituted by the color pattern of the books now on the third shelf is formed. Similarly, a new pattern is formed on the second shelf, and the old one on the third shelf is destroyed.

Not only are integration and disintegration involved in every change that takes place in the world, but also every occurrence of either is accompanied by both emergence and submergence. Thus, whenever any complex whatever is formed, emergence occurs because this complex has at least the character of containing just the elements it does, and this is an emergent character since none of the elements has it; and similarly submergence occurs because each element has lost the character it formerly had of not being a member of this complex. Again, when this complex is disintegrated, the characters of the complex which were emergents will become submergents, and the characters of the elements which were submergents will become emergents.

5. For emergence to be of philosophical use it would need to be non-ubiquitous and certainly detectable, but it is neither of these. In spite of the fact that emergence and submergence, integrational and disintegrational, accompany every change that takes place in

the universe, it remains very difficult to tell in any particular case whether a given character is an emergent or not or a submergent or not. Thus, to take the case of integrational emergents as an example, although we may be assured that all complexes have emergent characters, we can seldom if ever be sure that any given character of a complex is an emergent. For, in order to be sure that a given character X is an emergent of the complex R(A, B) we must make sure that X is not a character of a proper part of this complex. If we could be certain that A and B in the relation R constitute the ultimate elements of this complex, then perhaps we could ascertain whether or not they had the character X, and thus obtain an answer to our problem. But we can never be certain about this, because our analysis is always likely to be erroneous or incomplete. The complex may always be further analyzable, and it may be that X is a character of one of these minuter parts not yet distinguished. Thus, for example, before it was known that tannin was one of the ingredients of tea it might well have been thought that the peculiar properties now attributed to tannin were emergent characters of the tea leaves in hot water. When it is found however that tannin also is one of the elements of tea it is then discovered that these effects are properties of this element and not emergent properties of the complex. Similarly, some characters of chemical compounds which might have been considered emergents so long as the compounds were only analyzed into their atomic constituents may now be discovered to be properties of the electronic constituents of some of these atoms. So in the case of any complex, it is always possible that any character thought to be an emergent from it may be a character of some as yet undiscovered element of the complex.

6. These two facts, the ubiquity of emergence, and the difficulty of ever being sure that a particular character is an emergent of a particular complex, render the concept almost useless philosophically. For example, the very common inference from emergence to evolution is invalid. Emergence, as the manner in which novelty is introduced into the world, may be *necessary* to evolution, but it is certainly not *sufficient*. Emergence and submergence occur with every change whether evolutionary or devolutionary. Whether, as Alexander believes, the world has grown up from

elementary space-time to the complexity of God, or whether, as Plotinus asserted, it has unfolded from the unity and perfection of God to the multiplicity and imperfection of matter, emergence and submergence have both occurred at every step of the way.

The attempt of a number of philosophers, Lloyd Morgan, Alexander, Broad, Conger, H. C. Brown, and others, to erect a hierarchy of levels of emergents seems similarly not to be justified by the facts. The occurrence of submergents and of disintegrational emergents, quite neglected by these writers, makes such a scheme quite impossible. But even if only integrational emergents are considered the plan will not work. For, if complexes and their emergents are said to be on a higher level than the elements of these complexes, then there will be not a small and finite number of levels but a virtual if not an actual infinity of them. Simple entities will form one level, complexes of simple entities another, and so on indefinitely. Then there will be complexes made up of elements of different complexity. A theory of levels will be as complicated as the logical theory of types. For instance, it might be asked: What level of complexity is a Ford on? It is a complex composed of an engine, a chassis, and a body, and has properties which none of these have. But each of these parts in turn is a complex, the engine particularly consisting of parts which are themselves complexes having parts which themselves have parts and so on. There is such a tremendous number of levels and so many different series of them that the concept becomes almost useless. The distinction in any particular case between the levels of a complex and its elements is always important, for it reminds us that the complex has characters which its elements do not; but to arrange a hierarchy of levels is an impossible task.

What has been done by emergence philosophers is arbitrarily to pick out those levels which particularly interested them, and arrange them in the form of a hierarchy. These writers can be asked at once, of course, why *these* levels rather than others. It is remarkable that they nearly all disagree as to the number and nature of these levels. Thus Lloyd Morgan distinguishes three, or counting God four; Alexander, six; Conger, twenty-five; and H. C. Brown, five. Such disagreement itself indicates the futility of the labor.   The truth seems to be that any number of levels may be picked out according to the taste and inclination of the chooser.

To suppose further that such a hierarchy, if frameable, corresponds to the order of evolutionary development is simply to make an unwarranted leap.   Evolution may proceed a considerable way along one line of advance and scarcely at all along other lines.   It may then cease in the first line and develop rapidly at a different point.   This halting uneven advance cannot be attributed to emergence, for emergence takes place whenever complexes are formed, which is whenever a change occurs.   A slow evolution in some line or a temporary halt cannot be due to emergence alone, but must be attributed to an absence of those changes which bring about emergent characters that are evolutionary in their nature. Something in addition to emergence must be used to explain either a rapid or a slow or a null evolution, or a rapid or a slow or a null devolution.   Emergence alone is not sufficient to these.

Not only does emergence fail to imply evolution, but there is no necessary connection between emergence and value.   When a complex is integrated some characters emerge, but others submerge. Whether the value of those gained is greater than the value of those lost is a question which must be decided empirically in each case, as must also be that of the relative values gained and lost in the disintegration of other complexes which automatically accompanies the integration of the given one.   And finally, the net gain or loss of value in the integration of the given complex must be compared, also empirically, with the net gain or loss in the disintegration of other complexes which it necessarily involves. Clearly, from the mere knowledge of the occurrence of emergence no information can be extracted as to whether the world is growing better or worse.   Emergence is too ubiquitous to make possible any inference from it to anything non-ubiquitous, such as evolution or positive value.   Some emergents have positive value, others negative; some mark evolutionary advances, others retreats.

7. The attempt to solve some of the traditional problems of philosophy by means of the concept of emergence is no more successful than the attempt to make it imply evolution or value.   The view that it is useful in solving these problems seems to result from a confusion.   The concept of emergence gives a name and

thus calls attention to a commonly overlooked but nevertheless ubiquitous fact of the universe, the fact, namely, that some of the characters of every complex are different from the characters of any of the elements of that complex. From this it follows: (*a*) that no metaphysical theory which denies the occurrence of novelty, in the sense of denying that a complex may have characters its elements do not have, can be true; and (*b*) that some of the entities in the universe, the status of which philosophers have had difficulty in determining, may be emergents. These things follow from the concept of emergence and the fact of its ubiquity. Many things, closely associated with these, have been thought to follow, but do not.

(*a*) Although the fact of emergence destroys any metaphysical view which denies it, it supports no one of the remaining *Weltanschauungen* rather than any other. All metaphysical views which lay a claim to truth must be so modified as to include in their scope this newly recognized fact, but just because all views will admit it, it will help to validate no one view.

(*b*) Though the concept of emergence suggests a status which some objects in the universe may occupy, neither this concept nor the fact of the ubiquitous occurrence of cases of it proves that any given thing is an emergent. The difficulty of this latter task was pointed out in section five. Whether or not anything is in fact an emergent of any complex is in each case a problem for detailed empirical investigation. The method to be followed is illustrated by the work of the *Gestalt* psychologists. They have maintained that the concept of emergence is applicable to many psychological phenomena, but they have not let the matter rest there in the belief that they had solved many important psychological problems. Instead, in every case where they suspected that certain characters were emergents from a complex, they tried to establish or disprove this hypothesis by careful empirical investigation and experimentation. In the case of each such hypothesis what must be asked is: (1) What precisely is the character which is said to be emergent? (2) What precisely is the complex of which it is said to be an emergent character? (3) What are the ultimate elements of this complex? (4) Is it a matter of fact or not that the character in question is a character of this complex and is not a character of any proper part of it?   Only when all these questions have been answered favorably can we be justified in asserting of anything that it is an emergent character of such and such a complex.   Before they are answered such an assertion is merely a wild guess.

Consideration of a few concrete cases of the attempted application of the concept of emergence to the traditional problems of philosophy will bring out better than further abstract argument the things the concept does do and those it does not.   Take, for instance, the problem of the status of secondary qualities.   The concept of emergence suggests to Lovejoy [10] that secondary qualities are neither subjective nor objective, neither characters of the mind nor of the object perceived, but are emergent characters of the complex, the-object-being-perceived-by-a-mind.   But it also suggests other views.   Thus a New Realist like Alexander urges that secondary qualities are emergent from certain states of matter, that they occurred after primary qualities, but are nevertheless not mind-dependent because mind is not an element of the complex from which they emerged.   Thus the concept of emergence does not solve the problem of the status of secondary qualities at all.   It suggests a number of new solutions without throwing out any of the traditional ones.   Which, if any, of these are true it does not tell.

Lloyd Morgan argues that the fact of emergence shows that the New Realists are clearly wrong in asserting that nothing new accrues to an object when it is perceived.   For, he says, when it enters into perception the world is enriched by the emergent characters which spring from this complex.   On analysis, however, this conclusion of Lloyd Morgan's seems quite invalid.   For, though it is true that an object perceived is a complex and has certain characters which its elements do not have, this does not mean that any new characters accrue to the object which is one of the elements of the complex, or that it is altered in any way, which is all that the New Realist need concern himself to assert. The world in which the object is perceived is enriched by the emergents from this complex.   The complex has new characters, but

10 References to citations from here on are given in the notes to the first paragraph.

this does not mean that the object, an element of the complex, is enriched or has acquired new characters.

The mind-body relation is another problem which emergence has often been thought to solve. Thus, for example, it opens up the possibility of an Emergent Materialism, that mind, though different from matter, is an emergent from certain highly complex material states. But the task of making possible such a view and the task of proving it true are not the same. Emergence has accomplished the first task but it cannot accomplish the second. This can be done if at all only by the painstaking method suggested above.

Again, so far as concerns the concept of emergence or the fact that every complex has emergent characters, these are equally compatible with any view of the mind-body relation, whether it be parallelism, interactionism, or epiphenomenalism. They are equally compatible with Idealism, Materialism, or any form of Realism. For if everything that is, is matter, then there would still be emergents from every complex, but all complexes and all emergents would be material. The situation would be parallel but opposite if Idealism were true. And if Realism were true, both material and non-material emergents would be possible.

Emergence does not prove any view of the mind-body relation, but it can fit into all views. What it does do, however, is to make possible new and very suggestive theories which may turn out to be true.

The situation in regard to the mechanistic-vitalistic controversy is very much the same. Emergence has done the great service of making possible an intermediate view between the two old rivals, a view which tries to avoid the difficulties of each. This view asserts that living bodies do have characters that non-living bodies do not, and these characters require laws not exemplified in the inorganic realm. In this way the view attempts to guard itself against the stumbling block of the mechanists. Further, that the new characters which life has are due to some mysterious unobservable entity such as an entelechy or *élan* is denied by this view, and thus it is hoped the pitfall of vitalism is escaped. These peculiar characters are due not to an extra non-material element in living bodies, but to the particular formation of the material constituents of such bodies. The characters which distinguish living from non-living bodies are emergents from the complex of material elements making up such bodies.

Such is the view which the concept of emergence makes possible. That it is the correct view has not yet been shown. It cannot be shown merely by the concept of emergence or by the fact that all complexes have emergent characters. For though this last implies that the complex of the materials in a living body has emergent characters it does not imply that these are the characters which are distinctive of life.

Similarly, in regard to the problem of determinism or free will, the concept of emergence opens up new and suggestive theories, but it neither establishes the truth of these new views nor the falsity of any of the old ones. Emergence is compatible for instance with determinism, meaning by determinism the view that the occurrence of any entity Y is coverable by a law of the form, if X, Y, *i.e.,* every event Y is determined by some event X, for the occurrence of emergents is also coverable by laws, *e.g.,* if complex R(A, B), then emergent Y. On the other hand emergence is compatible with some kinds of freedom, such as the freedom of uniqueness and freedom of self-determination. For if there are, and of course emergence does not show whether there are or not, any acts of the human will which are unique emergents from the complex which constitutes a human being, then these acts are free in these senses. The concept of emergence is compatible with almost any view of this problem; it suggests new views but it neither proves nor disproves any view.

A few words ought to be said about the relation of the emergence theory to certain Idealistic doctrines. Two of these may be taken as typical examples, the notion of the Absolute, and the notion of a Society with its own Social Will, Social Consciousness, and so on.

That everything in the universe forms a complex and that this complex has emergent characters is a consequence of the emergence theory that is admitted. But that this implies that the universe is the Absolute is denied. For the Absolute of the Idealists has many definite characters, and that these are necessarily the same as the emergent characters of the complex which is the

25

universe remains to be shown. This might be true as a matter of fact but its truth does not follow *a priori*. It must be demonstrated empirically if it can be demonstrated at all, and in our present state of inadequate knowledge about the universe as a whole, the Absolutist's conclusion seems very precarious to say the least.

Similar is the situation with regard to the notion that Society is an Over-Individual with its own Will and Emotions and so on. The emergence theory implies that every group of people forms a complex that has emergent characters, but it does not imply that these characters are sufficient in number and kind to constitute the group an Over-Individual, and to make it reasonable to speak of the Will of the Group, the Mind of the Group, and so on. The theory of emergence makes possible such Idealistic views but it does not show them to be true.

The concept of emergence, then, has philosophic value in pointing out a fact which no theory may deny and in making possible new and suggestive answers to many of the standard philosophical problems. What must be insisted upon, however, is that it does not serve as a sort of magical formula by which metaphysical problems can be at once solved. To say that life is an emergent does not solve the mechanism-vitalism problem, nor does the assertion that mind is an emergent solve the mind-body problem. If such an assertion as that mind is an emergent from matter is to be even meaningful, it is necessary to state precisely not only what emergence means, but also just what characters the term mind is meant to denote and from just what complex or complexes these characters are asserted to be emergent; then, to know whether this now meaningful assertion is true or false, very painstaking experimental investigations must be pursued. The concept of emergence is a key which opens new doors to philosophic inquiry, some of which may lead to treasure, but it is not a master key which of itself unlocks the many doors of that seemingly impregnable castle where lie concealed the answers to the various problems of philosophy.

CHARLES A. BAYLIS.

BROWN UNIVERSITY.

4. A form of logic suited for biology

Walter M. Elsasser

Originally published as Elsasser, W. M. (1981). "A form of logic suited for biology," in R. Rosen (ed.), *Progress in Theoretical Biology*, Volume 6, ISBN 0125431066, Academic Press, pp. 23-62. Reproduced by kind permission of Elsevier. We are very grateful to Aloisius Louie and Linda Henderson for providing high quality digital scans of the original article.

The organized complexity of living organisms:
Walter Elsasser's contributions to complexity theory
A nonlinear life and career

The work of the German/American physicist turned theoretical biologist Walter Elsasser (1904-1991) is unfortunately little known today even though he made important discoveries in several scientific fields and played a key role in introducing the notion of *organized complexity*. Elsasser's conceptualization of this idea paralleled its elaboration by Warren Weaver after WWII (see Weaver's article "Science and Complexity" in issue 6.3 of *E:CO*). We at *E:CO* would like to start remedying this lack of recognition of Elsasser's seminal contributions to complexity theory by offering one of his key papers on organized complexity in biology. We're not doing this merely out of a desire to fill-out the historical record. Rather, we also believe that a careful reading and contemplation of Elsasser's main ideas can yield a sense of both how the sciences of complex systems have developed and where they might grow in the future.

The pattern of the life and career of Elsasser can itself best be appreciated as a nonlinear complex system because of its many unexpected twists and turns as well as the ample creative adaptations it took. Elsasser went from significant theorizing in atomic physics (including electron scattering and the shell structure of atomic nuclei), geophysics (specifically magnetohydrodynamics), atmospheric physics, solid-state physics, and then later turned his attention to theoretical biology. Along the way he became the recipient of numerous prestigious scientific awards.

In his memoirs of Elsasser, Harry Rubin (1995), points out that Walter had attended the University of Göttingen where he came into contact with such eminent physicists as Max Born, Paul Dirac, and J. Robert Oppenheimer. Göttingen was also a major center for mathematics being the academic home of the great German mathematician David Hilbert. This gave Walter the opportunity to also meet and form life long associations with two of the giants of twentieth century mathematics, Norbert Wiener and John von Neumann (who was Hilbert's assistant), both of whom played key roles in the development of the system sciences of cybernetics, information theory, and computer science. In fact, Rubin says that Elsasser's encounter with von Neumann's *The mathematical foundations of quantum mechanics* helped lead Walter to the conclusion that the introduction of probabilities into physics, a distinctive feature of quantum mechanics, did not so much 'loosen up' the framework of the theory, but made it even more deterministic and thus more suitable for reducing everything to physics than Newtonian mechanics had ever been. Prof. Elsasser grew to become even more deeply dissatisfied with reductionism, especially as applied to organisms. It took some twenty years of this struggle with quantum mechanics for Elsasser to propose a radical disjunct between the infinite sets underlying mathematical descriptions at the basis of physics with the finite sets of observations that the experience of biological organisms offered. Organisms, according to Elsasser, took the form of a heterogeneity or individuality which set the organic life apart from the inorganic world.

Studying later in Berlin, Munich, Leiden, Paris, New York, Ann Arbor, Chicago, and Los Angeles, Walter also became friends with other great scientists of the time including: Eugene Wigner, Leo Szilard, Erwin Schrödinger, the chemist and philosopher of science Michael Polanyi (who was himself an espouser of emergentism), Werner Heisenberg, H. A. Lorentz, Paul Ehrenfest, Wolfgang Pauli, Hermann Weyl, I. I. Rabi, Harold Urey, Enrico Fermi, Arthur Compton, and Robert Millikan. This is quite a list! And that was just in physics not to the mention all the other eminent scientists and mathematicians Elsasser became friends over his long and fruitful career.

Physics vs. biology: From homogeneity to heterogeneity

Emergentists of what I have elsewhere called the "mid-phase" period (Goldstein, 1999) suggested that if complexly organized living organisms displayed a building-up rather than tearing down of order, as they obviously and incontrovertibly did, there must be something going on with this complexity that obviated a simple transposition of classical thermodynamics into the realm of biology. Elsasser admitted that while on the surface it did appear life abrogated the Second Law, this surface look was based on an overly simple way of understanding biological organization. To replace this overly simplistic point of view, Elsasser offered his own version of an organicist philosophy of science which characterized the complexity of organisms in terms of a radical *in*homogeneity instead of the homogeneity found in how subatomic particles like protons or electrons were understood. That is, the latter were all identical in whatever substance they could be found. It was this homogeneity that allowed for the use of purely deductive probability rules in the statistical mechanical interpretation of entropy increase.

In contrast, Elsasser defined something he called "intrinsic" and "irreducible logical complexity" in terms of (see Elsasser, 1966; also see Polanyi, 1958):

1. an asymmetry between mechanistic and autonomous components - this corresponded to the asymmetry between the macro-variables and the immense reservoir of micro-states;

2. an integration or wholeness "whose origin in physics he has every right to consider as being rather obscure" - this integration could be understood in terms of the relationship of the individual to its class (discussed in his article below);

3. the fact that complex structures contained elements that unlike in physics were different from each other - biology dealt with the individual - even the classes grouping these individuals were heterogenous. As the eminent French historian of philosophy and science, Etienne Gilson (1966), pointed out, for Elsasser *inhomogeneous individuality* could be considered a metric of how complexity increased up the biological hierarchy.

According to Elsasser it was the unique individuality of a life form and its parts which demanded that their "regularities can be neither proved nor disproved on the basis of the laws of physics." Whereas physics and chemistry could, in principle, explain every detail in the functioning of an organism, they could not explain its existence. Although his distinction between existence and functioning was not particular illuminating - given the fact that in the case of an emergent organism its unique type of existence is its unique type of functioning as Whitehead had pointed out in his *Process and reality* - Elsasser did offer the very important emergentist suggestion that an adequate account of the living must take into consideration the manner in which life will always prove elusive to explanations couched entirely in terms of physics and chemistry.

Four principles of Elsasser's approach to complexity

We can summarize Elsasser's contribution to theoretical biology by appealing to Rubin's description of four fundamental principles in Elsasser's approach:

"The first principle is ordered heterogeneity. *Combinatorial analysis shows that the number of structural arrangements of atoms in a cell is immense; that is, much greater that 10^{100}, a number that is itself much larger than the number of elementary particles in the universe (10^{80}). But biology shows us there is regularity in the large where there is heterogeneity in the small, hence order above heterogeneity. This concept of ordered heterogeneity was first introduced by the molecular biologist Rollin Hotchkiss, systematized by the embryologist Paul Weiss, but given quantitative definition and set in a general theory by Walter.*

The second principle is creative selection. *A choice is made in nature among the immense number of possible patterns inferred in the first principle. The availability of such a choice is considered the basic and irreplaceable criterion of holistic or non-mechanistic biology. The term 'creative' refers to phenomena that, like everything in biology, are compatible with the laws of physics but are not uniquely determined by them. No mechanism can be specified by whose operation those selected differ from those not selected. He points out that the number of different patterns is also immense in the physical science of statistical mechanics, but in that case the variation of structure from pattern to pattern averages out. The patterns of inorganic systems repeat themselves over and over again ad infinitum, while those of each organism are unique...*

The third principle is holistic memory. *It provides the criterion for choice not expressed in the second principle. That criterion is information stability. The term 'memory' in a generalized sense indicates stability of information in time or, as in the case of heredity, the reproduction of information in an empirical sense, that is, without our knowing the full mechanism of reproduction. The creative selection of the second principle means the organism has many more states to choose from than are actually needed. The third principle says the organism uses this freedom to create a pattern that resembles earlier patterns. Walter borrowed the term 'memory without storage' from the philosopher Henri Bergson...*

Holistic memory requires a* fourth principle, operative symbolism, *to indicate that a material carrier of information is needed, namely*

DNA, but this acts as a releaser or operative symbol for the capacity of the whole organism to reconstruct a complete message that characterizes the adult of the next generation... In other words, operative symbolism is not necessary to the development of the postulational system of the first three principles that can do away with the conceptual difficulties and internal contradictions that always appear in any purely mechanistic interpretation of organic life. The informational system of organisms is therefore postulated to be dualistic; on one level it is mechanistic in the operation of the genetic code; on the other level it is holistic, involving the entire cell or organism."

Jeffrey A. Goldstein

References

Elssasser, W. (1966). *Atom and organism: A new approach to theoretical biology*, Princeton, CA: Princeton Univ Press, ISBN 0691079153.

Gilson, E. (1966). *From Aristotle to Darwin and back again: A journey in final causality, species, and evolution*, J. Lyon (trans.), Notre Dame: University of Notre Dame Press, ISBN 0268009678 (1984)

Goldstein, J. (1999.) "Emergence as a construct: History and issues," *Emergence: Complexity Issues in Organization and Management*, ISSN 1521-3250, 1(1): 49-62.

Polanyi, M. (1958). *Personal knowledge: Towards a post-critical philosophy*, London, UK: Routledge & Kegan Paul, ISBN 0226672883 (1974).

Rubin, J. (1995). *Walter M. Elsasser: 1904-1991: A biographical memoir*, National Academy Press, available online as *Biographical Memoirs Walter M. Elsasser March 20, 1904*, http://www.nap.edu/html/biomems/welsasser.html.

PROGRESS IN THEORETICAL BIOLOGY, VOLUME 6

A Form of Logic Suited for Biology

Walter M. Elsasser

Department of Earth and Planetary Sciences
The Johns Hopkins University
Baltimore, Maryland

I. Introduction: Biology versus Physics

A comparison of two sciences might appear as a rather primitive exercise. It is introduced here only because there are still many who insist on the purely empirical character of biology, i.e., the absence of a meaningful "theory" of biological phenomena. If a person enters the subject matter of biology coming from physical science, as this writer did, he cannot help but be intensely aware of the difference in character of the two sciences. It soon becomes clear that what may be called the hierarchical structure of physics has little or no counterpart in biology. By hierarchical structure we mean the possibility, which normally exists in physics, of condensing a more or less extensive area of experience into one formal statement, usually a set of differential equations from which, by mathematical methods, the description of those phenomena can be deductively derived. Nothing of the sort has ever been found in biology except in those limiting cases of physiology where the behavior of living things reduces to an application of traditional principles of physics.

Since physics deals primarily with extension, its chief tool of description is the *continuum*. Biology, on the other hand, starts with taxonomy;

23

correspondingly, the dominant concept in biology is that of a *class.* The difference between the two terms here italicized is found indicative of the distinction between the methods of physics and those of biology. Historically, the gap between those two concepts began to narrow when in the later nineteenth century mathematicians resolved continua into *point sets* (as a rule infinite); this gave rise to a logic of such sets (set theory). Now by their definition, which goes back to Euclid, points do not have an internal structure. When in the early years of this century Whitehead and Russell succeeded in combining logic and mathematics into one edifice, modern mathematics "took off." It is almost entirely based on sets whose elements are assumed to have no internal structure. This makes modern mathematics ideally suited for dealing with the constituents of matter discovered by the physicist. It is an experimentally well-established fact that those constituents, electrons, protons, and so forth are indistinguishable; their properties are such as to make these particles rigorously identical with each other. The more refined experiments allow one to specify this identity quantitatively to many decimal places. It is a well-established principle of physics that when one forms a class of, say, electrons, all elements of that class are strictly indistinguishable; it is as a matter of principle impossible to "label" the members of such a class so as to distinguish them individually. We shall speak of classes with this property as *perfectly homogeneous classes.*

Living things, on the other hand, are composed of many molecules. On leaving aside the limiting case of viruses for which the question of whether they are really living is not yet fully decided, apart from this, organisms can safely be assumed to be individually different from one another. This will be discussed in much more detail later on. Let us now only say that in all higher organisms this fact can be verified by direct observational inspection. If we assume the same to hold for lower, unicellular organisms, we are making a plausible generalization that is not contradicted by any known experience; in fact the observationally established adaptability of even primitive organisms makes the assumption that the elements of any class of organism are somewhat different from one another an extremely likely one.

Classes in which the elements are individually different from each other will be designated as *heterogeneous classes.* Thus, the description of biological states and processes is carried through in terms of heterogeneous classes. Ordinary logic pays little attention to the heterogeneity of classes. The basic abstract operations usually performed on classes (the junction and intersection of two classes) are of little interest in biology. There are two basic operations on heterogeneous classes which will be discussed later on; let me here just state them:

(1)   The selection of subclasses of a given class, such as to be of "greater homogeneity" than the original class. (Putting the term in quotation marks indicates merely that it is used in an intuitive manner and no quantitative measure of homogeneity has yet been defined.)

(2)   The inverse operation of selection will be designated as *embedding;* this is the construction (or demonstration of existence) of a larger class such that the class originally given is a subclass of this larger class.

It is found that a great many problems of "theoretical biology" can be tackled by means of formal operations applied to heterogeneous classes. Their clarification has, however, developed but slowly. (This is at least the impression of the present writer, who started as a theoretical physicist and later in his life became interested in applying the physicist's more abstract methods of analysis to the data of biology.) In such a clarification one is soon confronted with the well-known fact that few things in the scientist's world are more sterile and hence also more boring than sheer, abstract methodology. Methods, therefore, should only be developed in the context of a more concrete inquiry for which they constitute the tools. This author has in the course of the last quarter century written three books (Elsasser, 1958, 1966, 1975) which deal with an approach to theoretical biology that differs from others by a greater emphasis on the modes of thought of the physicist. This work is sufficiently specific to have, hopefully, avoided the pitfall of pure methodology. In the present review I have extracted and condensed those aspects of my method that deal with the adaptation of ordinary logic to the special requirements of biological science.

For nearly a century now, physicists have been imbued by a mode of thought designated as "positivism." My above-quoted three books may also be thought of as an effort at applying the physicist's positivistic mode of thought to the empirical material of biology. Here, we shall be interested in those aspects of positivism that refer to logic. Let me remark that it is certainly more than a coincidence that Aristotle, the founding father of scientific biology, was also the founding father of logic. It should then not be too surprising that an inquiry into matters biological turns into a discussion of logic.

But one finds here one of the worst cases of intellectual confusion that have occurred in the history of human thought. The term "positivism" was created by the Frenchman Auguste Comte (best known as the founder of sociology). It appears in the title of his work, *Course of Positive Philosophy,* which came out in the 1830s. From here Ernst Mach, who initiated the analysis of the scientific concepts used in physics, borrowed the term positivism. From this, there came also the tendency to think of

26   WALTER M. ELSASSER

this type of analysis as a form of philosophy. This arose out of an earlier
stage of thought when science grew up under the tutelage of philosophy as
was the case in the late Middle Ages and even later. But dragging in
"philosophy" puts science into an unnecessary, only historically founded
straightjacket.

Since the connection of positivism with traditional philosophy is so
often reiterated in the literature, we propose here a way by which the
scientist can try to dissociate himself from this connection: at first simply
a different terminology. We propose the term *structuralism* to replace
positivism. We begin by leaving everything except the terminology un-
changed. We are here primarily interested in extending the method of
structuralism ("positivism") from physics to biology so that it comprises
all natural science. In advance of this extension we shall describe in a few
words the way in which structuralist analysis is used to improve the
precision of scientific thought in modern physics. This recourse to physics
will give us occasion to consider the nature of structuralism and the general principles on which it is based. Only after this has been done shall we
apply a similar type of reasoning to some abstract aspects of biology,
which appear here in the form of a logical theory of heterogeneous
classes.

II. The Structuralist Method

As just announced, we begin our presentation of structuralism in
physics, the science where this mode of thought and analysis was born.
We do possess a quite unusually clear exposition of the method by one of
the great men of science, the astronomer Arthur Eddington (1939). The
title *The Philosophy of Physical Science* together with the book's appearance at a critical juncture of political history are probably the reasons that
the book has not received even a small fraction of the attention it deserves. What we readily notice is an implicit pointing to the past ("philosophy") rather than an outlook upon coming developments. Eddington
speaks of the positivistic interpretation of the philosophy of science; we
shall here speak of an independent foundation of physics by the use of the
structuralist method, both being the same except for terminology.

The central idea of Eddington is that of an abstract structure. This is a
scheme of mathematics in which certain symbols are defined, not by
pointing to objects of our external experience but solely by mutual,
equally abstract relationships between component symbols. Thus, for instance, in arithmetic the operations of addition and multiplication define a
set of number symbols. The most ancient abstract structures known are

Euclidean geometry and arithmetic (called algebra by mathematicians). For most of recorded history these two structures were considered unique. The famous philosopher Kant as late as the second half of the eighteenth century declared that space was a "form" of human perception, given "*a priori,*" as he called it. This is not really very different from the idea expressed by Newton a century earlier, to the effect that space and time are the "sensorium of God."

A radical revolution occurred in this mode of thought in the nineteenth century when mathematicians discovered that there was not just one geometry, the Euclidean, but many geometries differing from each other by their underlying axioms. Similarly, they found that there is not just one algebra, the one taught to children, but a vast variety of abstract algebras differing again in their axioms. Mathematics at present knows a nearly inexhaustible plethora of purely abstract constructs of this type.

Some of these abstract structures can be used to represent the physicist's observations. I presume that most of my readers are acquainted with the fact that the theory of general relativity uses four-dimensional curved spaces. Quantum mechanics in its turn uses, instead of conventional school algebra, a noncommutative form of algebra combined with certain other abstractions (Hilbert spaces and Hermitean forms) that play a role in higher mathematics.

Now the structuralist method of theoretical physics consists in this that certain properly chosen abstract structures are used as images of a body of observational data. No philosophical or theological preconception enters this process because the abstract structures are in the first place defined so as to be independent of any human experience that would go beyond constructions of pure mathematics. Eddington points out a number of implications of this method, of which I will mention here only two. First, one may try to define the aim of scientific theory within the framework of such a method. As Eddington emphasizes, the principal aim of any scientific theory of physics is not to "explain" phenomena but simply to describe them. The concept of explanation as used in physical science can have two different meanings. First, it can indicate model making that tells us how a certain natural process works; for example, one can model the operation of a geyser. Second, explanation often has a purely logical meaning. This occurs when an observed regularity is recognized as a special case of a more general law, for instance, a specific chemical reaction as the application of a law of chemistry. But by conceiving of the analysis of some natural phenomenon as the application of an abstract structure to fit a given situation, one makes clear the principal function of theory as an imaging process, that is, as a description of observed reality. Psychologically speaking, there is no possibility of an involvement of the pupil's person as there is in the less precise concept of science's providing "explanations," where assent is essential.

One benefit of the structuralist method mentioned by Eddington is the disappearance of the term "existence" from the vocabulary of the scientist. All propositions of science are essentially relational; thus, while all kinds of propositions saying that objects A, B, and C interact with each other can and will occur, no statement occurs that "A exists" except in the sense that A is subject to these interactions. Eddington describes the term "existence" as a "metaphysical" predicate and tells us that the structuralist (his positivist) method is a powerful tool to separate science from philosophy, the latter in its more extreme form of metaphysics.

One may believe, however, that it is possible to go one step farther. Implicit in the structuralist method there is the hope that the scientist (for the time being only the physicist) can liberate himself from an affliction that has beset the more thoughtful part of humanity since the beginning of time: the tyranny of words, which seems to pervade the history of human thought. There is good evidence that in much earlier times some words had magical power—one need only think of the taboo on pronouncing the name of the diety so conspicuous in the early parts of the Bible. In more modern times there occur arguments about "understanding," but this understanding is often a semiconscious process involving some unknown depths of the mind. As against this, the structuralist analysis of the laws of modern physics realizes understanding in terms of a large number of small steps whereby the elements of an abstract structure are related to numbers of observations, thus minimizing (so far only for physics) that slipperiness of concepts which seems such a spectacular characteristic of all learned discourse (even of much scientific discourse) of the past.

Such insights would be of little value if the structuralist method would forever be confined to physics. We may next define a task which this author first formulated some years ago: the structuralist method that can be taken to define scientific clarity and absence of prejudices ought to be expanded beyond physics proper, into biology. The pages to follow should be understood in this sense: they are meant to be an application of the structuralist method to matters biological, primarily to an analysis of the modifications of this method which are required to make the transition from physics to biology.

For the purposes of this task, before we can even begin to enter into any technical detail, we must focus on the overall, qualitative differences between the empirical material of biology and that of physics. There has always existed, in physics and its direct applications such as astronomy, a prevalence of mathematical methods; the material is such that it lends itself to mathematical analysis. But the relations between the biologist and the mathematician has never been very close; often they have degenerated into downright hostility. In the nineteenth century, for instance, biological journals simply refused to accept articles that contained mathematical formulas. Such incidents were not quaint exceptions but expressed the posture of the biological community during much of its history. What is it that gives the biologist an attitude so radically different from that of the physicist?

To shorten the path one has to traverse in order to arrive at a clear-cut statement of this difference, let me give at once the result: while the material of the physicist lends itself as a rule to displays of regularity that can be expressed quantitatively, the material of the biologist is characterized by a pervasive complexity and variability. We shall fasten on the term *complexity* as expressing in a condensed form the chief characteristic of the biologist's material. Needless to say, the mere term stands here for a whole history of experiences which, properly speaking, ought to fill a book; we must here simplify to keep the size of this review within reason.

One of the remarkable aspects of the complexity of living things is that its full extent has been discovered only in comparatively recent times, as history goes. The microscope was invented shortly before the year 1700, and further tools for the elucidation of biological complexity, the ultramicroscope, X-ray analysis, the electron microscope appeared only much more recently. If, as Nordenskiöld (1946) has put it, organic life constitutes "a separate form of matter," then the chief characteristic of this form of matter can be clearly enunciated. It consists of the presence of structure within structure within structure, as a distinguished physiologist has put it. This distinguishes a living object clearly from an inanimate one, say a rock. The rock is structured on two levels, that of its microcrystallites, and again at the level of its molecules. But the living tissue is, as a general rule, structured at *all* levels. Whenever new instruments make accessible new domains of observation, new forms and dimensions of structure appear. This ubiquity of structure is a basic property of living things, and one that affords a rather clear-cut distinction between living and inanimate matter. We may think of this pervasive structuration as the detailed expression of the complexity which we just declared a basic property of organic life.

What, then, is the theoretical description of the complexity and this structuration? When we go back to ordinary, scientifically unrefined discourse, we realize that the members of a logical class do not yet have the property of being equivalent to each other. The class of all cows does not imply that any two cows are substitutable for each other. Instead, the description of cows depends on so many variables that the cowherd finds it always possible to tell the animals apart. This property is readily recog-

30   WALTER M. ELSASSER

nized as a general one for all the logical classes that appear in biology. Correspondingly, we shall say that the biologist employs heterogeneous classes. This, then, is the point at which we propose to separate biology from physics: Biology will be taken as using a logic of heterogeneous classes while physics employs homogeneous classes. The following pages are devoted to an inquiry into the properties of heterogeneous classes. We hope to show that this can be taken as a transfer to biology of the structuralist methods that have been so successful in the development of modern theoretical physics.

A main point of difference between homogeneous and heterogeneous logical classes lies in the use of mathematics within a universe of homogeneous classes. Those, for instance, who are more closely acquainted with the mathematical theory of quantum mechanics know that the entire theory of the homopolar bond centers around the equivalence and interchangeability of electrons and could not even exist without this trait.

Quantum mechanics offers us a prime example of mathematically rigorous laws of nature. What interests us here is the opposition between the idea of quantitative regularity as expressed in a "law of nature" and the complexity that we recognized as the chief characteristic of organic life. On closer view, the notion of a law of nature is found to have two implications: there is first the notion of quantities that are uniquely determined by law, and there is second the notion of quantities that vary from case to case and that we shall describe as contingent variables. (They are more familiar to the physicist in the form of initial conditions and boundary conditions.) It may not always be possible to separate clearly, in a quantitative sense, these two ingredients, namely, dynamical variables and contingent variables, but their logical distinction is clear enough. That is, the distinction between general laws and contingent effects may be dubious in individual instances, but when it comes to classes of processes the distinction can readily be made evident.

Now let me recall that according to the previous arguments physics is the study of the laws of nature whereas biology was so far only characterized as a realm of utter complexity. The following question then arises: Does this complexity lead to the need for basic conceptual innovations, or can this complexity be fully understood in terms of the intricate molecular mechanism which experience has shown to exist in all living beings? If one adheres to the latter assumption, one can be said to have adopted a physicalistic approach. The meaning of the term physicalistic is precisely that no major conceptual changes are assumed to occur on going from physics into biology. The more common term *reductionist* will here often be used interchangeably with physicalistic. "Reduction" is a rather general term for the logical operation of subsuming one regularity under a more general one. Here we are, of course, concerned with the relationship of biological order to physical law, a special case of reduction.

In the practice of biological and especially biochemical research the physicalistic approach has given rise to so formidable successes that I need not go into them at this place, assuming that my readers know a great deal already about the unraveling of the "genetic code" and other accomplishments of "molecular biology." But there remains a persistent question: Can the distinction between living and inanimate matter be exhausted entirely in terms of the familiar type of chemical mechanisms? Or what else may be necessary to account for that "particular form of matter" known as living things?

We meet here, however, an entirely novel situation as far as the progress of theoretical science is concerned. In the past, when a conceptual innovation became necessary (and it did indeed a number of times in physics), this led in all cases to quantitative predictions about the results of the "new" theory which differed from those of the "old" theory. Having made the critical measurements, the scientist could abandon the old theory and replace it by the new one. But this simple criterion does not apply to the problem that concerns us here. The fact is that there is no shred of evidence anywhere in the vast literature of biochemistry or biophysics that the laws of physics (in practice the laws of quantum mechanics) are invalid or stand in need of any modification. Any approach to theoretical biology must start from this basic fact. But at the same time this does not absolve us from the need to pursue the possibility of a substantial conceptual innovation occurring in the passage from theoretical physics to theoretical biology. In other language, any conceptual innovation must be such that it leaves the laws of quantum mechanics invariant. Of course, all molecular physics and chemistry, including all biophysics and biochemistry is assumed derivable from quantum-mechanical theory.

We have already indicated the key concept around which such a conceptual innovation will center; it is complexity, with its attendant variability. This complexity can be rendered in logic by the heterogeneity of classes, and it is the heterogeneity of classes, then, that will serve as the chief vehicle to set off theoretical biology as against theoretical physics. We shall later on have occasion to see how this heterogeneity of classes can be understood in such a way that it does not violate the laws of physics. We shall, in particular, deal with irreducible heterogeneity, that is, with heterogeneity such that no logical operations can be found which would allow one to resolve heterogeneous classes into a combination of homogeneous classes. For the time being I can just claim that this is possible; the specific theoretical arguments will appear as we proceed. A main argument will be that the heterogeneous classes of the biologist can be thought of as subclasses of the homogeneous classes of physics which transcribe the physicist's "laws." Under these conditions, as we shall see, the question of whether heterogeneous classes can be considered the formal equivalent of some "extra" set of laws, need never arise.

My claim is then that the formalization of complexity through the irreducible heterogeneity of logical classes (as defined later) introduces an element into biology which represents conceptual innovation and so goes beyond the reductionist approach. Since at the same time this can be done without violating the laws of physics, we have a situation that, as already pointed out, seems to have no counterpart in previous "conceptual revolutions" known to the theoretical physicist.

The arguments given here have been gradually developed by this author and appear in the three books already cited that span an interval of over 20 years. I have here presented them in a novel and, I believe, clearer fashion. Still, my aim has not altered: to explore the potential for a conceptual innovation germane to biology, that conforms to the laws of physics.

III. Finiteness. Individuality

The main concept we discussed so far, that of a heterogeneous class, fits the structuralist method: it is purely abstract and free from philosophical connotations. And while homogeneous classes are the equivalent of a mathematical treatment, heterogeneous classes do not lend themselves easily to mathematical representation, the less so the more heterogeneous the classes envisaged. But even if there is little of a mathematical structure, we can still speak of a *formal* treatment. We have indicated already that such a formal treatment in terms of heterogeneous classes might be the starting point of a genuine theoretical biology; but if we ignore mathematics as a structuring element, we have little to draw upon for the ordering of the empirical data of biology.

We can proceed farther by realizing that benefit may be drawn by concentrating upon properties which enhance the consequences of heterogeneity. The results now to be presented show that it is indeed meaningful to speak of an enhancement of heterogeneity; we may take it as a heuristic program that the abstract structure of heterogeneous classes should be enhanced in the direction of bringing out heterogeneity by such logical methods or technical tricks as we can find.

Homogeneous classes yield to mathematical treatment because their members differ little or not at all from each other. Let me exemplify: hydrogen atoms which are all in the same quantum state form a perfectly homogeneous class. If the individual hydrogen atoms are in different quantum states but if the chemical nature of the atoms is always that of hydrogen, we are still dealing with a homogeneous class but the homogeneity is no longer perfect. The number of members in such a class has little significance. This is so, precisely because the logical operation of formation of the class can be replaced by a mathematical description of its members, in which all the members of the class correspond to one and the same mathematical sumbol. This fact is deeply embedded in the language of the physicist: he speaks of a "system" which he represents by a mathematical symbol. He fails to specify whether his mathematical symbol represents one object of his experience (one atom) or a whole set of atoms or perhaps even an infinite set.

On closer scrutiny, the last-named alternative turns out to be of theoretical significance: Should one or should one not admit classes of *infinite* membership into the logic of description? Note that for a perfectly homogeneous class this question has no meaning for, by definition, the members of such a class are indistinguishable from each other. Hence, there is no operational meaning to the distinction between finite and infinite membership of the class. We have here a nearly classical case of the use of the term operational meaning. The early (positivistic) students of the meaning of scientific terms in physics carried out the first such analysis; on this basis others in the early years of this century constructed first the theory of relativity, then the quantum theory. We now find a subject of logic, namely the finiteness of classes, that promises to play a similar role with respect to theoretical biology as certain concepts of space, time, and causality had played earlier with respect to the structuralist reconstruction of physics proper.

It is obvious that the number of operations which may be carried out in the real world is finite; in other words an infinite set of operations, or else a class with infinite membership, are idealizations. We cannot know in advance, of course, whether a postulate excluding classes of infinite membership will have far-reaching consequences. In the event that it does, as we shall see, we shall from now on postulate that all classes used in biological descriptions have finite membership. We shall not spend time in justification of this postulate in advance of its application; we expect that the reader who continues to follow the argument will come to recognize the remarkable degree of novelty that this postulate allows us to introduce into the formal analysis of biological conditions.

The finiteness of biological classes may appear less strange when one remembers that according to the testimony of astronomers our whole physical universe is finite in its extension in space and time. This was first stated by the astronomer Hubble in the 1920s and has since come to be acknowledged as a basic astronomical fact. We do not need it here, in the sense that we can just as well think of the finiteness of classes as a matter of biological method. But even so, it may become more palatable by the reference to the astronomers' universe.

A remarkable benefit of the use of finite classes lies in the fact that this allows one to deal in a formal way with the concept of individuality. Let me lead to this discussion by some introductory words.

Owing to the historical descent of science from philosophy, many scientists have never abandoned the belief that scientific terms arise out of words of the language; and so if a word exists, they believe that there exists not just a concept but, potentially, a scientific concept. Structuralism is the sieve that more critical scientists have put in the flow of their experiences to keep out the verbalizations in which ordinary language is so rich. The structuralist method tells us that science deals only with observable relationships, where the language of ordinary intercourse deals with an unanalyzed jumble of facts, fancies, and wishful thinking. Here we find that one of the principal logical tools which screen the language in a structuralist sense is the distinction between infinite and finite classes.

It seems impossible to define the term individuality in a universe of discourse consisting of homogeneous classes, where the distinction between a class of finite and one of infinite membership cannot be given an operational meaning. Consider, then, heterogeneous classes of finite size. Assume the number of members in the class has not been specified; it may be small. We can now define an individual as a heterogeneous class with only one member, a one class. The question of how meaningful and valuable such a definition of individuality is cannot be decided by purely formal arguments; it requires the application of the concept of heterogeneous classes to (biological) experience in order to see how well an object for which individuality is claimed can be logically separated from other objects. Such operational isolation of individuals turns out to be a matter for the empiricist, the taxonomist.

Now if one considers individuality from the viewpoint of the empirical scientist, it becomes soon clear that this property tends to enhance the difference between the physicist's way and the biologist's way of looking at nature. For the physicist, who is as a rule concerned with homogeneous classes, the individual differences between one specimen and another specimen of a class (e.g., of a class of crystals) are incidental, and the physicist tends to ignore them except when the individual features form by themselves a class (e.g., the class of semiconductor crystals containing a certain proportion of impurities dissolved). But the biologist does not think of individuality in this way. The field biologist in particular tends to think of every (higher) organism that he meets in his field work as an individual different from all other organisms of the same species.

To make clearer the benefits that accrue from finite classes consider the common method of the selection of subclasses in order to demonstrate the presence of a suspected mechanism. Given the high degree of structural complexity prevailing in living things, there may arise the case that the class has been exhausted before a decision about the presence or absence of a mechanism has been made. In other words, in a universe of finite classes there exist questions that cannot be decided operationally. This can, of course, only occur in a world of great complexity; in such a world, the running out of specimens indicates the possibility of posing unanswerable questions. In such a world, therefore, the term *irreducible complexity* takes on a precise operational meaning. It is the irreducible complexity defined in terms of an exhaustion of finite classes that makes a physicalistic biology unsatisfactory.

This is the central statement of the theoretical scheme proposed in these pages and followed up previously by the author, in his books already quoted. We shall study some of the implications of a theory of heterogeneous and finite biological classes and see in particular what the relationship of these classes is to the homogeneous and therefore in practice unlimited classes of the physicist (unlimited reproducibility of experiments).

Lest the distinction of heterogeneous as against homogeneous classes be taken as a scholastic exercise that is of little avail to the practitioner of biology, we shall next discuss some observational results that are closely related to the heterogeneity of biological classes. This will give us the first occasion, in the course of a so far rather abstract inquiry, to connect our abstractions more closely with observation. In biology there has in the past been no clearly recognized nonphysicalistic concept which could be compared with experience. Now the heterogeneity of classes is a concept characteristic of biology; as we mentioned already, it is to be enhanced and enlarged but it will not be supplanted later by other nonphysicalistic concepts. To make what is to follow more comprehensible, let me remark that individuality is a limiting case of heterogeneity, which implies that each member of a class has characteristics that distinguish it clearly from any other member. In the practice of the naturalist the two terms are often used interchangeably; in what follows we shall now and then adopt the habit of saying individuality if we mean heterogeneity of classes. This is not likely to lead to a misconstruction of meaning. On the other hand, I believe that we are faced here with a situation where the famous statement of the philosopher Hegel applies: when quantitative differnces become extreme, we are inclined to perceive them as qualitative differences. This

36   WALTER M. ELSASSER

may be taken as our guiding idea in investigating the difference between heterogeneous and homogeneous classes; we shall try to show that the observer often perceives this difference as that between the living and the inanimate.

Almost a quarter-century ago there appeared a little book entitled *Biochemical Individuality* whose author is the biochemist, Roger J. Williams (1956). By that time Williams had acquired an extensive reputation as a biochemist, through his work on the identification and analysis of vitamins. This book is based on a collection of empirical observations, some gathered from the literature, many others obtained in Williams' own laboratory at the University of Texas. There is a simple thesis to the book. It is this: every organism can be distinguished from every other organism of the same species through measurable characteristics that are different in any two of them. (The quantitative observations are so far limited to higher organisms, but the impression this reviewer has gained is that this is a technical rather than a fundamental limitation; in other words, we have found no good argument against the assumption that heterogeneity of classes is a basic property of *all* living things, not just of higher organisms; this assumption is implicit in our later general discussion, and the existing gap should be filled out as observational techniques continue to improve.)

Williams' book is a short one (209 pages), but it contains a vast amount of information, amost all in the form of factual reports of quantitative data. It is unusual to find oneself confronted with so novel a doctrine in such a purely practical guise. There is no substitute for the reader's perusing this book (which is readily available in the trade), but we must indicate at least in outline in what the novelty of Williams' approach consists. Williams succeeds in shattering the concept of normalcy. Since any higher organism is clearly characterized by a very large number of variables, any definition of a ''normal'' organism must comprise the need for specifying these parameters within ''reasonable'' limits. But what does reasonable mean? Williams' results demonstrate with perfect clarity what it does *not* mean: the ''normal'' is not something in the nature of an average that can be determined observationally to be confined to be within certain percentage limits in a statistical sense. Such a definition would be suitable for an engineer who deals with products coming off an assembly line. The engineer could, for instance, specify that any value, x, of a variable which deviates by less than 10%, say, from an agreed-upon mean value, x_0, is acceptable, whereas any values of x that deviate from x_0 by more than $\pm 10\%$ must be rejected. Williams' abundant data show that a statistical dispersion of the order of 10% is totally inadequate to characterize any kind of ''normalcy'' in biological investigations; there are innumerable parameters that will vary within a population by factors of two or three, going as high as factors eight or ten or even higher.

Williams, before he enters into the main body of his data, which deal with analyses of biochemistry, devotes a chapter to anatomical variations. By means of drawings taken from a well-known textbook of human anatomy he demonstrates the vast variety of shapes that occur, for instance, in human stomachs; one finds that the volume of the stomach may vary by a factor of eight among humans that in medical tests would be declared perfectly "normal."

Williams is, of course, mainly interested in variations of a biochemical nature, and his book is full to the bursting with examples. To save the reader a multitude of details I will confine myself to reproduce the summary of his Chapter 5 entitled "Individual Enzyme Patterns" (p. 77).

> *Summary.* The cumulative evidence that each individual human being has a distinctive pattern of enzyme efficiencies is hard to refute on any rational basis. Furthermore, inter-individual variations in enzyme efficiencies in normal individuals, insofar as they have been determined, are not of the order of 20 to 50 percent, but are more often at least 3- to 4-fold. Differences of 10- to 50-fold (!) have been observed in a substantial number of cases even when the number of normal individuals tested was small.
>
> Certainly these differences are far from trivial. Even to the author, who has been interested in variability for some years, the extent of the variability comes as a surprise. He therefore cannot blame his colleagues if they seem incredulous. We have included in our discussion every enzyme for which we have found substantial data, and the least inter-individual variation we know of appears to be about 2-fold.
>
> Inter-individual differences related to metabolism come to light only when *detailed items* are compared. When two individuals of the same height and weight yield total metabolism values that are about the same, it is easy to conclude that their metabolisms are substantially identical. The evidence presented in this chapter, however, indicates that the details of metabolism in two such individuals may be very different indeed. The extent to which specific reactions may take place may vary 10-fold! This idea is admittedly difficult to accept, but it appears to be substantiated by concrete and cumulative evidence.

It would be impossible, in a report like this, to do justice to the wealth of data presented by Williams. We shall not even try to do so but refer the reader to the original work, which can be read without difficulty. Certainly, the ideology of the biomedical profession will have to undergo a radical change as Williams' data and method become more widely known. Since the data are factual, they cannot of course be ignored, but neither have they so far been recognized and taught as they so obviously deserve. The present writer, who cannot by any stretch of the imagination be thought of as a member of the biomedical establishment, can do no more than present this impressive empirical correlate of the more abstract reasoning discussed in our pages; we hope that scientists will eventually yield to the persuasion of an overwhelming array of factual data. It appears

38   WALTER M. ELSASSER

from Williams' book that the physicalistic view of biology, according to which one would expect classes of organisms to correspond to classes of relatively uniform mechanisms that can be built up on an assembly line, is factually wrong. This result must appear to be more the beginning of a new and unaccustomed line of research than a simple fact which could be integrated into any existing point of view about the nature of organic life. For the time being we can think of the results of Williams' work as the observational counterpart of our abstract concept of heterogeneous classes.

Let me come back once more to a point of our theoretical arguments, namely, the relationship of the heterogeneity of biological classes to their finiteness. We have above introduced the finiteness of classes by way of a postulate, that is, abstractly, without reference to empirical observations. In Williams' work there is implied, however, another possibility which he does not state with brute clarity but which will not escape the diligent reader: the vast observed differences in concentration of certain compounds as between different individuals suggest that the *metabolic patterns,* the underlying feedback cycles, may differ from one individual to the next. Such a variation of metabolism seems to be confirmed by certain observations: the concentration of a substance is found to be much less variable if tested in one and the same individual at different points in time than when one compares different individuals with each other. Given the tremendous variety of feedback loops already observed in metabolism, it must seem mainly a matter of established modes of thinking, whereby investigators are kept from testing the constancy versus variability of metabolic feedback pattern in given classes of organism. Such a task has a purely mathematical (statistical) side to it in addition to its empiricist aspects; we shall come back to the mathematical aspects a little later.

One is dealing here with a type of relationship between theory and observation that is well known in physical science but in the past has remained all but unknown in biology. For science to be at its most fruitful, it is advantageous that its theoretical and its observational components be of comparable degrees of difficulty. While this generally holds in the physical sciences, it has not usually been true in biology, where often either the theoretical aspects or the observational aspects were trivial. When one deals with the heterogeneity of biological classes that are also finite the mathematical theory (combinatorics) of such abstract structures becomes rather complicated, and so do the statistical observations on the extended samples required for such work. But there is little to indicate that they cannot be mastered by modern techniques.

The biological experience that is contained in the preceding statements, first implicitly in the discussion of heterogeneity and then explicitly by way of Williams' results, may be used as the starting point of an obvious generalization: each living thing has a measurable individuality. This cannot at the present time be proved. It may be thought of as a horizon which delimits biology. This writer finds it useful to think in such terms, because in this way we can define theoretical biology in a purely formal manner. If such an idea can be carried through in detail, we remain fully within the purview of a theory that conforms with the demands of the structuralist method. We should then be free of the entanglements of speculative philosophy so far as biology as a separate science is concerned.

The remaining pages are given to an overview of the formal problems that are raised by such a program. The program, to say it once more, consists of the assumption that each living thing, each organism has measurable features that allow one to distinguish it from any other organism. On continuing this inquiry we shall find that in such a scheme the main question centers on the relationship between the physicist's description that uses mathematics, logically expressible in terms of homogeneous classes, and the heterogeneity and individuality of the biological description. The magnitude of this problem makes one recognize the inadequacy of any one investigator; but it seems clear that the problem is rationally posed and hence can be dealt with, ultimately, by a rational analysis, even though the technical difficulties might be very large indeed. But one main aspect of this formal approach is that it is likely to lead us into abstract problems of a rather formidable, purely mathematical complexity. This concerns answers to the question of how the universe of heterogeneous, finite classes, which is that of biology, can be quantitatively related to the universe of mathematical structures, transcribed into a logic and mathematics of homogeneous classes, as given to us by atomic and molecular physics.

Leaving these questions for a later section, let me draw attention to one very attractive feature of this formal scheme: it limits science intrinsically. According to the structuralist view, the purpose of theoretical science is to put relational order into the observed phenomena. If the principal abstract tool used is logic, then this amounts to the establishment of classes. Such classes are homogeneous in the case of physics, heterogeneous in the case of biology. The limiting case of a universe containing only different individuals as just envisaged corresponds to a bound for science that lies in the nature of science itself, not in any external convention. In our age which, as one can daily read in the newspapers, is the great age of science, it is gratifying that one can find at least one major branch of science which carries its own limitations built into it in a natural way. Using a somewhat antiquated language we might say that the order which appears in biology through the existence of classes shades off into chaos in the form of an assembly of distinguishable individuals.

40   WALTER M. ELSASSER

In the sequel, when we speak of biological classes we shall consistently assume that these classes are heterogeneous as well as finite and that they can be broken down into a set of distinguishable individuals by suitable observations.

IV. Finite Classes and Selection

In the preceding pages we have developed a formal scheme for the representation of biological data. This turned out to be a variant of ordinary logic, namely a logic of heterogeneous, finite classes. We found empirical grounds for thinking that this is a very good way of describing a large part of all biological data, evidenced by the amazing observational results of Roger Williams. But is this really enough to characterize that "separate form of matter" which consists of living bodies? In such an estimate, the theoretician (such as this writer) does tend to exaggerate the power of specific tools, the abstractions and formal operations. We must, in addition to abstractions, have more facts drawn from experience that allow us to set living matter off from inanimate one. This section is devoted to the enunciation of some such facts; others will be given in Section V.

As I have tried to make clear in the beginning of this article, I am approaching biology here with the mode of thought as well as with the techniques of the theoretical physicist. The outcome of the preceding inquiry was that the *logic* of the two sciences is different: physics uses homogeneous classes (which lend themselves readily to mathematical formulations), whereas the preferred tools of the biologist are heterogeneous, and finite, classes. Admitting this, we are faced with the question: Can anything specific be said about the relationship of the actual classes used by the physicist for his description and those actual classes that are used by the biologist? We shall see that a general proposition can be enunciated and that it expresses the difference between the living and the nonliving state in a more substantive manner than would be possible with the preceding, purely formal constructions.

It is significant that, while this distinction between the living and the inanimate is not a consequence of any logical formalism, it need, nevertheless, be expressed in terms of a logic of classes. In simpler words: the distinction between the living and the dead is a matter of classes, not of an individual event. This is so much a part of everyday human experience that we are usually not even aware of it. Whether a person is alive or dead at a given moment is decided by criteria that are based on *general* experience with the behavior of members of the species (e.g., the significance of heartbeat). In physics the distinction between one kind of interaction (e.g., electromagnetic) and another (say, gravitational) is much simpler because the corresponding classes are homogeneous; telling the dynamical variables apart from the contingent ones (boundary conditions, etc.) offers little difficulty.

The general tool of description of the physicist when he deals with bodies containing many atoms or molecules, is statistical mechanics. This branch of physics predates the advent of quantum mechanics by about half a century. After the discovery of quantum mechanics these two branches of physics were found to coalesce readily into a unified mode of description known as *quantum statistical mechanics*. Expositions of this subject are available in numerous textbooks. I cannot, of course, indulge in an exposition or even in a précis of the content of this particular branch of science. What I can and shall do is to extract some general conclusions that are sufficiently simple and unambiguous that they are (1) intelligible to a broader, biological audience and (2) acceptable to the overwhelming majority of the specialists.

We shall remember first that statistical mechanics, being a branch of physics, begins by dealing with homogeneous objects, that is, those which contain only one or two, or at least an extremely small variety of molecules. For such assemblies of equal molecules one replaces the study of their complicated internal dynamics by a study of statistical distribution functions. In connection with these distribution functions it is useful (and is often so done in practice) to replace all continuous distributions by *discrete* ones, that is, to replace the continuum by a set of small intervals. The general term for such intervals in multidimensional space is *cells*. Such a subdivision is not only applied to ordinary space but also to velocity (or momentum) space. (In general, however, when we speak in this report of "cells," we use the term in its familiar biological meaning.) Since we are interested only in objects which occupy a finite volume and have a finite internal energy, the discretization implies that all numbers which appear in the calculus of distribution functions are finite. This fact will be very useful if presently we apply this type of reasoning to biological objects.

A standard mathematical question that arises in statistical mechanics is this: In how many ways can one distribute m objects (molecules, thought of as indistinguishable from each other) over n empty cells? The answer is given by the binomial coefficient, that is, this number is

$$\binom{n}{m} = n!/[m!(n - m)!]$$

where the *factorial function* is defined by the multiple product,

$$n! = 1, 2, 3, \ldots, n$$

of all integers from 1 to n. In any textbook of mathematical analysis one can find Stirling's approximation to the factorial function for large n,

$$n! \sim (n/e)^n$$

We note here the appearance of the expression n^n which represents a rate of growth with increasing n faster than the well known exponential growth, e^n; we shall call it *factorial growth*.

Now the mathematical details just given are not necessary for us in detail. What we must consider is the formidable magnitude of the numbers generated by factorial growth. The objects to be counted in statistical mechanics are numbers of atoms or molecules; exceedingly large numbers will appear all the time as the result of factorial growth. We can write the factorial function

$$n! = n^n e^{-n} = (10^{\log n})^n e^{-n} = (10^{n \log n}) e^{-n}$$

so that factorial growth means a growth such that $n \log n$ appears in the exponent. Consider now that n stands for the number of molecules or else for the number of places available for molecules; it is clear that this kind of mathematics leads to tremendously large numbers, a fact familiar to many students of physics, since statistical mechanics is a part of the advanced curriculum in physics.

We shall now express the facts just mentioned in an alternate fashion. Let me introduce a limit that separates "ordinary" large numbers from extravagantly large ones. Let me take arbitrarily the number 10^{100} as boundary line; I call any number "immense" that is larger than 10^{100}. Conversely, if a number is smaller than the reciprocal of this number, 10^{-100}, I call it *immensely small*. The operational implications of this obviously somewhat arbitrary definition can readily be made clear: immense numbers of operations, however simple they may be, cannot be carried out in the real world. (Computer theorists call them "transcomputational.") To appreciate this one has only to remember that astronomers commonly give an estimate of the size of the universe by saying that it contains about 10^{80} atomic nuclei. We are fairly sure that the lifetime of the universe expressed in seconds of time is less than 10^{18}; thus, 10^{100} may be taken as an upper bound for the total number of events (1 sec apart) that can occur in the real world. Clearly, the precise choice of such numbers is not as important as the insight that there exist numbers so large that they no longer correspond to any set of events in the real world; if such numbers are imagined as physical events, they can represent thought experiments only.

We can now readily make contact again with the structuralist mode of thought that was introduced early in this review. The scientist imbued

A FORM OF LOGIC SUITED FOR BIOLOGY   43

with the structuralist method will be set to wondering when he encounters numbers that cannot be realized operationally. He will recognize a challenge for the reconstruction of the conceptual scheme of science along structuralist lines. This is the kind of argument that will lead us to a critical evaluation of the dividing line between biology and physics. But so far in this section we have spoken only of the objects of the physicist. They form homogeneous classes because they have a homogeneous physical constitution, that is, they consist only of one kind or of a very minute variety of distinguishable molecules. But living things form heterogeneous classes because they have a heterogeneous constitution; they contain a much larger variety of different molecules as well as steric structures than one usually encounters in inorganic chemistry. If we want to find out about the difference of living and inanimate bodies, we should have to apply statistical mechanics to the heterogeneous bodies which constitute the object of the biologist's researches, as it has been so successfully applied to the homogeneous bodies of the physicist.

Here we have to admit that we are defeated before we even begin by the magnitude of such a task. None have gone before us: there is no statistical mechanics of heterogeneous objects. But we cannot afford to give up. We ask whether we cannot sketch some general ideas that may provide a lead in this vast jungle of complexity and heterogeneity. This author has proposed one such idea which he calls "the principle of finite classes." It appears in each of his above-quoted books; we proceed now to its exposition.

Before doing so it is useful to get a glimpse, in concrete and intuitive terms, of what the heterogeneity of structure of organic tissue really implies. In contradistinction to an inert inorganic body, such as a crystal, the living body metabolizes incessantly. The extraordinary complexity of the living organism is due in large part to the well-known ability of the carbon atom to form multiple bonds. If enough carbon atoms are assembled, they will form chemical structures in a great variety of ways. This variety is known to the organic chemist as the phenomenon of *isomerism*. To illustrate isomerism, take the very simplest case of only four atoms connected by three bonds. They can be arranged in two quite distinct ways, as shown in this diagram:

(a) ·—·—·—·   (b) ·—·—·
                              |
                              ·

In case (a) the four atoms are all in a straight line; in case (b) there is a side branch. Now if one has a large number of carbon atoms, one can build up a tremendous variety of structures by just combining connections (a) and (b) in various ways. All these edifices are legitimate structures within organic chemistry, usually with H, O, and N atoms interspersed. Any two

44   WALTER M. ELSASSER

carbon atoms that are close neighbors (say less than one angström unit apart) can be related in three ways: by a double bond, a single bond, and no bond at all. The application of a more quantitative analysis shows that if the number of atoms, especially carbon atoms, is thought of as growing large, the number of possible distinct chemical structures increases factorially. This is the result that we shall presently require. It is characteristic of organic tissue and would not be found to hold in any other material body. Using the specific terminology introduced a little while ago, we can say that the number of ways in which an organic body the size of, say, an ordinary cell can be realized chemically, is immense. Since the structures involved are those of ordinary chemistry, an observer with sufficiently powerful instruments should have no difficulty in principle (although the practical difficulties may be great) of ascertaining what this structure is in a concrete instance.

For any lump of matter having the size of a cell and its typical chemical constitution there exists thus an immense number of ways in which its detailed chemical structure can be realized. On the other hand, the number of cells of any given species existing in the real world cannot be immense. This is a simple result of all the estimates of the magnitude of this number (the number of cells of a given species) that one can carry out. This shows that there exists a gross disproportion in numbers between, on the one hand, the number of ways a given gross chemical constitution of a cell can in principle be realized, and, on the other hand, the number of specimens, the number of actual bodies that can appear in the world. The latter is vanishingly small, often immensely small, compared to the former. The enunciation of this disproportion will be called the principle of finite classes. Our next step will be to shed some light on its nature and implications.

In the first place, let me comment on the use of the word "principle." We do not have in mind here an axiom or a postulate but merely a very general result of a numerical estimate. Its value lies in its applicability to all organic matter. One of its chief advantages lies in its *crudeness*. We compare the size of a class, that of potential chemical constitutions, which is found immense, with a class of actual living objects, which while extravagantly large is not immense. We hence do not have to enter into any subtleties, all of which can be bypassed at this stage of the analysis. We also do not plan to write a book that would contain a more detailed "proof" of this principle. Such a book could, however, no doubt be written by a person who combines a thorough knowledge of organic physicochemistry with experience in statistical mechanics.

In order to proceed from here, we shall simply accept the "principle" as true and go on to evaluate its consequences. To repeat the chief result: of all the possible molecular patterns of a living object only a vanishingly small subclass appears in reality. If we formulate the principle in that way, we are at once led to the further question: By what "mechanism" or additional rule do organisms select the patterns that do appear in the real world? We are using here the term *pattern,* indicating a purely static case; if we would want to speak of chemical transformations, we would use the term *process* instead. The mathematical analysis used in statistical mechanics is closely similar in the two cases.

So far we have spoken of a form of quantitative analysis that pertains entirely to the chemistry and physics of highly complex "organic" compounds. This analysis involves, on closer view, the *probabilities* for the occurrence of the various chemical reactions, that is the reaction rates. But probabilities can be mathematically defined with any kind of rigor only as limits of infinite sequences in infinite sets (infinite classes). Since we have eliminated all infinite classes in favor of finite (although possibly immense) classes, probabilities are not here mathematically defined. However, one can always define relative frequencies in any finite class: one simply compares the (finite) number of times an event A occurs in the class with the number of times another event, B, occurs. I can here only hint at these deep-lying questions known to those who have studied the foundation of probability theory. In the present context we plan to focus on the point where our propositions differ from the more usual reductionist preconceptions, and we have now arrived at this point.

If we accept the principle of finite classes but do not proceed from there to reductionist or physicalistic assumptions the latter of which imply that everything which happens in the organism is the result of physical causality combined with "randomness," then we must consider the alternative. A selection of patterns takes place among all the possible patterns. We assume this selection to be a spontaneous natural phenomenon not subject to experimental control. Only a negligibly small subclass of all the physically possible class is so selected.

Since these assumptions are somewhat novel, they deserve further discussion. In the first place, the relationship of biological regularities to physical laws is here clearly defined. The classes of biology are *subclasses* of the classes used in the description of the physicist. Furthermore, experience indicates, as reported in Section III, that biological classes are heterogeneous, whereas we know that the order of physics must be represented by homogeneous classes. This alone suggests that the heterogeneous biological classes are not the equivalent of any law of nature; hence, conventional vitalism which postulates a modification or extension of the laws of physics differs strongly from this scheme.

If we ask about the nature of those patterns that are *not* selected, we can say in the first place that since they are of immense number, they cannot even be explicitly enumerated. Now we are dealing here with the description of a living thing down to and comprising the molecular scale. The patterns on that scale which are compatible with the laws of physics but which have *not* been selected may be described as *molecular terata*. (Most of my readers will know that the term "teraton" is used in medicine to designate a more or less misshapen product of embryonic development.)

The most characteristic feature of this theoretical view of organic life is that in contradistinction to reductionism it postulates a specific distinction in the pattern of behavior as between living and dead objects. In such a framework, we believe that our scheme has a greater degree of what in the English language is described as *verisimilitude*. In this word, the Latin root, *versus*, meaning true, is readily recognizable. A scientific theory must have some degree of verisimilitude, resemblance to the naively perceived truth about the observed facts, something conspicuously lacking in the methods of reductionism. We can express these ideas also in another and perhaps more readily comprehensible form.

If one crosses the borderline between physics and biology, the question arises as to whether this must involve a conceptual innovation. According to the view presented here, this is most emphatically the case. It contrasts with the widespread view that in crossing the boundary no significant conceptual innovation is required, the physicalistic view commonly described as reductionism. It does not suffice, of course, to say that such innovation is needed, one must deal with the problem by indicating in what this innovation consists. The first and most important point has already been introduced: a logical theory of classes including a scheme of heterogeneous classes is required. The classes of biology which our experience shows to be intrinsically heterogeneous appear here as subclasses, of vanishingly small extent, of the immense, homogeneous classes with which physics provides us through the application of statistical mechanics; however, the same approach gives heterogeneous classes of immense magnitude if applied to the particular structures of organic chemistry that are found in living things. In a way, therefore, the heterogeneity of biological classes is preformed in the facts of biochemistry.

It should be clear that in the presence of a major conceptual innovation I can give no more than a *program* for a novel theory, that it is too early to elaborate on the details of any such theory. My principal aim at this stage must be to bring out and exhibit as clearly as possible the logical contradictions that appear in the current practice of biology, and to show how the new formulation removes these contradictions. Also, the introduction of the new logical scheme increases greatly what I called a little while ago the verisimilitude of the theoretical scheme. It seems that in the complex field of theoretical biology a certain tradeoff is possible: simplicity of assumptions, especially retention of physical principles only, versus verisimilitude. I believe that by introducing the concept of a selection of biological classes which form a negligibly small subclass of physically possible patterns (the entire, immense class being one of terata), by doing this we have removed any logical contradiction between two simultaneous types of laws: the laws of physics and regularities of biology. Biological regularities, as we have indicated, do not derive from universal laws; since they result from a process of selection, they are *specializations* of the type of behavior derived from the laws of physics; they are therefore by definition nonuniversal.

The particular regularities underlying biology according to this scheme do not stand alone; in the next section we shall deal with a further type of phenomena that leads to a loss of verisimilitude in the usual reductionist treatment. But before doing this, let me have a glance backwards, upon a past way of dealing with the same or a closely similar problem. We are thinking now of the famous *Cartesian Method* introduced by Descartes in his *Discourse on the Method,* which appeared in 1637. In it Descartes describes his method of how to deal with complicated objects of research. Since most of the complex objects that a researcher encounters are living things or at least the products and results of organic life, the Cartesian Method has played a particularly great role in physiology. According to Descartes, one investigates a complex object by breaking it down into simpler components; one then studies these components one by one so as to grasp their functioning. At the end one puts the components together again, at least mentally, and in this way gains an idea of how the whole things works.

The notion of the Cartesian Method has survived to this day in many semipopular and semiphilosophical expositions. Nobody, however, can read Descartes' description without being reminded of the way a teenage boy takes a watch apart. This is quite natural. Historians tell us that the seventeenth century was the age when natural philosophers came to visit the workshop of the "artificer"; it was the golden age of mechanics where God was conceived as the designer of a "clockwork universe." The benefits that accrued to biology in this period are tremendous. For example, Descartes was a somewhat younger contemporary of William Harvey. And if we look at the Cartesian Method once more, closely, we can readily perceive the point at which it differs from the assumptions made here. In order that the component pieces of a complex object may be thoroughly investigated, one requires *repetitive* experiments; in other words, the components must be *homogeneous.* In the world of real biology, heterogeneity prevails at all levels; we would have to break organic tissue down into its constituent nuclei and electrons to achieve homogeneity. This is the reason the Cartesian Method does not apply beyond a purely macroscopic level, and in this sense, but in this sense only, there is a well-defined meaning to the statement that *biology is non-Cartesian science*.

Finally, we should point out the relation of our scheme of selection to the dualistic or quasi-dualistic schemes that have been surrounding biology for a long time. Outstanding among such ideas is that of Niels Bohr regarding "generalized complementarity" which he enunciated repeatedly in the 1930s, the years after the consolidation of quantum mechanics. Bohr's idea clearly was that there might be a phenomenon analogous to the wave–particle duality of quantum mechanics but at a higher level. It is hard to see, however, how such a scheme can be reconciled with our basic idea of selection where the biologically preferred class of patterns is negligibly small, or immensely small as compared to the size of the class comprising all patterns compatible with the laws of physics. To express this numerical relationship clearly, we raise its content to the level of an abstract principle: We introduce a *principle of selection*. It asserts in essence that living things become defined only by a selection being made by nature whereby the actually occurring states are distinguished from the immense multitude of possible ones. The precise nature of the selective process is here deliberately left open. The reason for the latter caution is that it seems unlikely that its specifics can be determined without extensive recourse to empirical evidence (especially of an embryological kind). Nevertheless, it seems to make sense to state the principle in the generality given it here.[1]

V. The Stability of Information

The postulate of selection just introduced is so far purely formal; it does not tell us by what criterion the selection is made. This selection has to do with the *morphology* of the organism. Since we are sure that there will be no contradiction with the laws of quantum mechanics, it is appropriate to consider in detail the idea of a conceptual innovation that does not violate quantum mechanics, for instance, in the following context.

There are frequent hints in the biological literature to the effect that the Second Law of Thermodynamics may be violated in the organism. These hints have, to my knowledge, never been further elaborated. The main

---

[1] In another paper (Elsasser, 1981), I have replaced the term selection by the more specific one, "creative selection." Reasons for this elaboration are given or, at least, sketched in that paper.

reason is that the authors of such remarks are confusing two similar things: the Second Law in its generality and the application of the Second Law to information theory. The conceptual novelty to be contemplated will have to do only with the application of the Second Law to information processing. The making of an assumption in this field requires in the first place that we define information. Even a layman can see that information presents itself to us as a sequence of symbols. Speech is a sequence of symbols—similarly in television transmission: a picture is resolved into a series of symbols that are sent over the air by electromagnetic waves.

Information theory deals with certain quantitative aspects that can be abstracted from such a sequence of symbols. For now we leave the connection between symbol sequences and morphology open; we shall specify it later as required. It is clear that a picture on a television screen may be thought of as a form of morphology, and we shall assume, as is customary in information theory, that there exists a process whereby this "morphological pattern" can be transformed into a sequence of symbols and eventually back again from the sequence of symbols into the morphological pattern.

To be specific, let us think of the symbols transmitted as letters of the alphabet. An important concept is that of an *error in transmission:* a letter of the message is by this error replaced by some other letter, in a random substitution. Modern information theory started when in the year 1948 the mathematician C. Shannon showed that out of any sufficiently long sequence of symbols one can construct a quantity, a number known as the *entropy,* which has the property that on substitution of new (erroneous) symbols into the message the magnitude of the entropy increases in an overwhelming number of cases. Put in terms of probability, on random substitution it is vastly more probable that the entropy increases than that it decreases.

One has to be careful to distinguish the traditional Second Law from the special application of its formalism to information sequences. Let me therefore designate the result that the entropy of an information sequence increases when errors are introduced as *Shannon's Rule.*

The formal apparatus that leads to Shannon's Rule is identical with the one which is used to derive the Second Law of Thermodynamics from statistical mechanics. But it is somewhat simplistic to think that therefore heat and information should be confounded. To give an example taken from the physicist's practice: the equation of heat conduction and the equation of diffusion are mathematically identical. But one cannot conclude from this that heat and a diffusing substance are otherwise similar. Here, we can conclude that the formalism is so well developed that one need not admit any question about the validity of Shannon's Rule. Now if we assume that any morphological feature can be translated into an information sequence, that is, a set of symbols, then the effect of Shannon's Rule as applied to random errors is the deterioration of morphological features as time goes on. This is still a very loose statement; but since by our previous postulate all admissible morphological features arise by a selection from the class of physically possible ones, neither the validity nor any assumed invalidity of Shannon's rule can lead to a violation of the laws of quantum mechanics.

In information theory the effects of random disturbances upon a message are described by the term "noise." There is ample evidence, which will be indicated in the remainder of this chapter, that in living things the effect of noise in degrading morphological order is found to be very much smaller than one would infer from Shannon's Rule. This experience will be summarized in the *postulate of information stability,* the second of the two postulates that set off biology from the physics of the inanimate. Its full implications will gradually appear as we go on. For now we note that on assuming Shannon's Rule not to be fulfilled in organisms, whereas it is in all electronic devices, we are giving a very specific meaning to the claim that organic life constitutes "a separate form of matter," as Nordenskiöld (1946) expressed it.

There are many ways in which the stability of information against deterioration, and often the reproduction of preexisting information, appear in biology, even to the untutored observer. Ordinary (cerebral) memory is a well-known special case of this stability; so is the phenomenon of healing, including, for instance, the well known ability of lower organisms to replace lost limbs. But the most important for the biologist is plain *heredity.* It has the advantage that here the scientific concept and the concept of everyday human life do very nearly coincide, so no fanciful refinements are necessary. Since I am now dealing with a difficult task, namely, the introduction of a basic conceptual novelty, I shall confine myself here to heredity as a most conspicuous example for the discussion of information stability in organisms. This does not mean that I am unaware of other cases such as those just mentioned; it merely means that in this review in place of being comprehensive, I shall confine myself to one phenomenon, heredity, which seems broad enough.

The scientific study of heredity has a history, of course, although a limited one. This history began about 300 years ago when the invention of the microscope enabled biologists to study for the first time the cellular aspects of the process of animal reproduction: fertilization of the ovum by the sperm and early embryonic development. Even without the example of computers before their eyes these pioneers of biology could pose the problem of heredity in terms of an information sequence that is transmitted from the parents to the progeny. "Information" being by definition a sequence of symbols, the question inevitably arose: What is the nature of the symbols transmitted in heredity? In the eighteenth century two schools of thought arose. The adherents of one view, called the theory of *preformation,* claimed that a small but fully adequate "model" of the adult organism is found in every germ cell. Then there was the opposite view to the effect that the germ cell was endowed with a "potential" to regenerate the adult in the absence of a full information sequence. This became known as the theory of *epigenesis.* Of these two only the term epigenesis has survived into present-day biology. In the course of history the term has acquired various kinds of subtle meanings which I will not discuss. Here I shall take the term epigenesis in its pristine meaning, namely, indicating a process in which the handling of information violates Shannon's Rule and hence cannot be described by a computer model.

This would seem to impute upon the earlier biologists a capacity that they did not have: the capacity to make a quantitative judgment as to what a computer can and cannot do. But I believe that this is not serious because even if a person does not yet have the tools to execute a given task in detail, he may well be able to judge reasonably well the limits of the possible.

We spoke of the postulate of information stability as giving a more specific content to the "selection" introduced in Section IV. We can now say with some precision what this postulate of information stability implies. It would tell us little that is new so long as the transmission of information in the processes of heredity can be modeled by a well-defined computer. It is when the effort at modeling by computer breaks down that the postulate of information stability enters to acquire its specific meaning. Now remember that what this postulate describes is the nature of the selection among the immense number of possible patterns that according to physics (statistical mechanics) are possible. The overwhelming majority of these are terata and are of little biological interest. Those which are selected are such that they tend to conserve the morphological information beyond what would be possible in a world of automata.

It should be apparent that in this way we attribute to the organism a quality best described by the term *creativity.* It should better be called recreativity since the patterns to be created follow those of preexisting patterns. Since in introducing selection we have deprived ourselves of an exhaustive recourse to physical causality, we are thrown back upon the concept of a heterogeneous class as embodying the formal aspects of the process of selection. Previously, the theory of heterogeneous classes may have appeared as a tautological reshuffling of accepted logical concepts. But used as a foundation of the postulate of information stability the heterogeneous class appears as an abstract but irreducible descriptive element of biology. It plays a role in biology similar to the role that differential equations play in physics. We do not think to ask what "spirit" induces the planet to move along the orbit prescribed for it by Newton. Similarly here, we do not ask what "vital agency" makes the selection; we merely describe what we see happening in nature.

The decisive fact, as will readily be perceived, is the one summarized by the principle of finite classes: the number of possible molecular patterns of, say, a cell, is immense and vastly exceeds the number of specimens of any appropriate biological class, so that the transition from physics to biology requires a selection of patterns. The selection that occurs at any one point in time is guided by the fact that the progeny is a member of the (heterogeneous) class of all ancestors.

(I believe that the reader will find it just as difficult as this writer has for many years to absorb the idea that "mere" membership in a heterogeneous class has results analogous to those of a "law of nature" in a more conventional sense. This indicates that the notion of universal laws, so dear to the seventeenth century, will perhaps not always remain the last word in the description of nature. But I am firmly convinced that the unwillingness to discuss conceptual innovation on the part of large numbers of reductionistically inclined biologists and biochemists is not due to a commitment to the philosophical doctrines of rationalism. The psychological connection is in the reverse direction: it lies in the erroneous belief that conceptual innovation must be connected with a change in *universal* laws that has kept the older philosophy alive. The central point of my scheme is that a quite radical conceptual innovation can occur without any violation of the laws of physics.)

Two kinds of questions now present themselves. First are those of method and general "philosophy"; I shall discuss those in Section VI. For the time being I shall concern myself in more detail with the *empirical basis* that underlies the postulate of information stability. I shall try to show that the empirical evidence points powerfully toward a postulate, such as the one sketched, as the expression of the radical conceptual innovation required to penetrate from physical science into biology.

To summarize, in order to deal with heredity at all, we first need a *model*, however crude. If it turns out to be false, we can correct it later; but a good model as a starting point is of immeasurable value in keeping one's ideas precise. The model we have in mind here is that of an electronic device in which signals circulate; for simplicity we shall speak of a computer, even if we deal only with the transmission of information, not with its modification. One quantitative effect that cannot be completely eliminated is that of noise; for instance, thermal noise is ever present. It can be reduced only by going to low absolute temperatures, a condition not of interest in biology. If in a message consisting of letter symbols we represent the action of noise by a substitution of "false" letters for the original ones, the message will become progressively less intelligible. Shannon's Rule then implies that this loss of intelligibility progresses always in the same direction. For this reason technical devices that minimize errors must be of prime importance to the engineer as well as to the biologist.

The engineer who designs computers achieves a minimum interference of noise through making the signal energy much larger than the (mean) noise energy. In ordinary computers, even miniaturized ones, signal-to-noise ratios of many millions to one are found routinely. It is apparent that at this point the comparison of organisms with computers may lead to useful results. One might expect that the organism also separates its "signal" from inevitable noise. But we run here into a severe limitation of the computer model as applied to biology. A computer has fixed design features and variable signals circulating in this fixed design. In the organism we have just metabolic activity that can only artificially be split into a stable and a variable part—there is only metabolism. Nevertheless, one will try to use a computer model when one is dealing with properties which are independent of a somewhat artificially introduced splitting into a "signal" effect and a general background. Certain observational facts concerning the metabolism of living tissue are so general that they allow us to do this.

In an earlier work (Elsasser, 1958) I have given a quotation from a then widely used college textbook of organic chemistry (Conant and Blatt, 1947, Chapter 20) bearing the name of a distinguished practitioner of the science. The passage summarizes the situation in an admirable way, and therefore I shall simply repeat it. Since I do not claim to be an organic chemist the correctness of the statement will have to be argued in any event by others than myself. The statement is:

> Biochemical reactions as a rule liberate or absorb relatively small amounts of energy; a balanced or nearly balanced equilibrium is often at hand. It seems that living cells operate with reversible reactions where possible and can utilize or absorb energy in only small amounts. Thus in the oxidation of carbohydrates a complex series of changes takes place so that at no one step is anything near 100 kg-cal of energy liberated, which would result if all at once one carbon atom of a carbohydrate were oxidized by air to carbon dioxide. Apparently this necessity for reversible reactions with relatively small energy changes is a characteristic of biochemical transformations (p. 376).

Nobody with even a humble background in physical chemistry could possibly mistake the meaning of this statement. If we interpret metabolic activity in terms of the ratio of signal to noise, as one must do if one wishes to apply a computer model, then one could restate it by saying that metabolism seems always to proceed very close to the noise level, and such a statement is quite independent of the details of any model adopted. There is then good evidence that whatever "signal" is attributed to metabolism, this signal is not properly separated from thermochemical noise. It means that such signal-processing abilities as we attribute to the living cell operate within or close to the noise level, not far above it as it should according to our understanding of information theory.

Let me now compare this biochemical result with the stability of morphological features as it can be observed in cases where heredity is of prime importance. The science of paleontology offers us occasion to observe the variation of morphological features over lengthy periods of time. Species, once they have become established, tend to undergo only relatively minor changes of their morphological features during their lifetime to extinction, which lifetime is usually of the order of several million years. Thus, the morphological features remain approximately constant over some millions or some hundreds of thousands of generations of the species. Again, this observational result is very general and applies to all kinds of fossils.

The contradiction between the biochemical result that reactions do not deviate far from equilibrium and so signals cannot be preserved from noise, and, on the other hand, the paleontological result concerning the long-term stability of morphological features—this contradiction is so glaring that I propose to describe it by a separate term. I shall speak of the paradox of heredity. There is, of course, nothing new about this paradox; it has been known for a long, long time. What is new here, if anything, is its explicit recognition as a paradox, that is the admission that such a paradox might not be capable of resolution within the physicalistic world that is built up from organic chemistry together with basic notions of computer science. We claim therefore that here is the occasion of a conceptual innovation. We have already enunciated its content: it is the postulate of information stability. This postulate defines what we usually call creativity and which more appropriately should be called *recreativity;* we spoke of this before and shall briefly come back to it in Section VI. For the time being we shall remain with the concrete aspects of, on the one hand, heredity, and, on the other, computer models that are meant to describe heredity transmission.

Let me try once more to explain, in slightly different terms, why it is that a loss of information by mixing with noise is so much more fatal in the case of morphological features than it is in the case of the molecular order–disorder relationship that underlies the Second Law, a case with which the order in a biomolecular information sequence is so often compared. In the inorganic case one has just as many molecules, but one measures only two or three "variables of state." Even when a condition of maximum entropy has been reached, one can usually, by a simple change of conditions, return to a state in which the entropy is no longer a maximum. In the case of morphology the number of parameters is as a rule large; this results in a condition where the number of possible patterns is immense in the above-defined technical sense of the term in the absence of the constraint represented by the information. Thus, information corresponds to a selection from an immense reservoir; once the information is lost, it cannot be retrieved otherwise than by actually going through this immense reservoir of variants, a procedure whose impossibility in the real world is clear enough. There exists no analog to this behavior in the physicalistic world with which we deal with the help of the Second Law.

Given the crucial importance of the preservation of information, it must long have been a challenge to the engineering mind to think up devices that serve for the protection of information sequences from deterioration. It is of significance, therefore, to see that practically nothing has been achieved in the 30 years since information theory came into existence and has occupied some of the best mathematical and engineering minds. The only method that has ever been proposed to protect information from loss is by *redundancy*, which is just a learned term for the repetition of the information. A specific device for using redundancy to maintain information has been proposed by J. von Neumann (1956). He calls it a "majority organ," and it seems worth discussing. One assumes that in place of one transmission line, computing device, or the like, one has three identical ones in parallel, all three carrying the same message. There is also inserted a device which at regular intervals compares the three messages with each other; when one of them differs from the other two, it changes the former so that all three are alike again. Quantitatively, if e is the (small) probability that one device makes an error, then the probability that this triple device makes an error is readily seen to be of order e^2. Variants of this device with more than three parallel transmission lines or computing devices can readily be conceived.

What is impressive, however, is the extreme clumsiness of such devices. The impression is very strong if not overwhelming that this state of affairs is not due to the relative newness of the engineering art involved but comes from our human inability to extract a message from an immense reservoir of competing ones if the message has once been lost in the reservoir. In the language introduced in connection with our first postulate, the selection of a "good" morphological pattern out of an immense set of terata is hopeless once failure has occurred.

As explained in the beginning of this section, we have chosen the phenomena of heredity to illustrate the application of the postulate of information stability. There are other applications, among which ordinary (usually called cerebral) memory is most important. A recent note of the author (Elsasser, 1979) contains some suggestions as to the form this postulate is likely to take when applied to cerebral function. We are in this case in totally unexplored territory! But such history of this problem as exists indicates that here, perhaps more than anywhere else, conceptual innovation is of the essence. In the article quoted, I have shown that the postulate of information stability, based again implicitly on the principle of finite classes, seems to represent the desired type of innovation.

VI. Selected Patterns

The introduction of membership in a heterogeneous class as criterion of the behavior of living things in place of mathematically expressed "laws" is a large conceptual novelty that requires more discussion. As I hope to show, we cannot as yet say in detail just what it implies but we can readily state certain things it does not imply.

Many of my readers will have wondered why I insisted so carefully—petulantly, they may have thought—on applying the structuralist method at every step. A little reflection will now show why this was done: at this juncture we cannot fail to encounter a well-known form of philosophy, namely, Platonism. The concept of an idea, or prototype, underlying every class will, on even superficial structuralist analysis, be found incapable of being expressed in terms of abstract relationships between observed data, the basic feature of structuralist science. Here, the scientist finds himself before a difficult choice: he can either consummate the marriage with Platonic metaphysics, whereupon the biologist seems forever committed to this hybridization of his science, or else he must explicitly dissociate biology from metaphysics by taking a strict positivistic-structuralist approach. In my own writings I have resolutely taken my stand on the latter alternative. I do not claim, however, that the alternative can be uniquely expressed by such terms as "true" or "false." The distinguished British biologist B. Goodwin (1978) in a recent article has clearly indicated that he takes the opposite position. He claims that some form of metaphysics, for which he chooses that of Alfred Whitehead, is necessary to interpret the data of embryology. Here, I shall not try to persuade my reader that my structuralist method is true as against a commixture with metaphysics. I shall be satisfied with having shown how closely together lie the problems of embryonic development and the concepts of Platonism.

The structuralist method allows one to make more explicit the distinction between a scientific analysis of heterogeneous classes and the traditional reasoning of philosophers. The philosophers dealing with these questions became later split into Platonists and Aristotelians. Those who derived their thinking from Plato called themselves "realists"; they claimed that the Platonic ideas were the only real things in the world and actual objects just more or less flawed copies. The opposite party called themselves "nominalists" since they believed that heterogeneous classes are mere "names" invented by logicians to put some order into the vast variety of phenomena. Now even a modest degree of analysis indicates that neither of these two traditional philosophical attitudes is acceptable from the structuralist viewpoint: realism suffers from the fatal flow that Platonic ideas are not operationably verifiable abstractions. But nominalism, on claiming that classes are no more than formal pigeonholes invented by logicians, runs counter to our assumption that membership in heterogeneous classes does have an operational meaning if applied, for instance, to heredity.

So we see that the problem of the operational meaning of membership in heterogeneous classes is wide open; it will no doubt require much further research, partly of a theoretical type, in a field where so far all efforts have been almost purely empirical. The status that we have claimed here for heterogeneous classes is that of a primary and irreducible type of natural order, on the same level as the more conventional "laws of nature" so familiar to everybody. It is clear that this new concept of regularity, as it was formalized in our two postulates, implies the autonomy of biology. If now we ask what the consequences of this autonomy are, the distinction of our approach from the well-established physicalistic–reductionistic one can be viewed much more clearly.

Our first password on entering biology from the side of physical science was complexity, this being the most distinguishing feature of all things living. We concluded, toward the end of Section IV, that, hence, the Cartesian Method does not apply in biology. This complexity was, so far, implicitly understood as phenomenal, that is, it refers to the anatomist's or physiologist's observation of a multitude of devices interrelated with each other in a variety of ways. The Cartesian Method had suggested that one should look for simplicity underlying the phenomenal complexity, a simplicity that can be unearthed when one breaks a complex object down into simpler (and therefore presumably more homogeneous) small parts. Now if we say that biology is non-Cartesian science, it becomes of interest to see whether to the phenomenal complexity directly observed there corresponds also a *logical complexity*. Assume there are two mutually incompatible mechanisms, or models, (a) and (b), each of which could by itself produce an observed effect. Then in our world of heterogeneous finite classes, we might run out of specimens before we have formed a class which corresponds exclusively to model (a) or exclusively to model (b).

Now the ancient controversy regarding the process of hereditary transmission, the argument of preformation versus epigenesis, lends itself to an interpretation of this type, or so we may assume by way of a working hypothesis. The discovery of the genetic code has given the physicalistic interpretation of heredity a tremendous boost. But our inability to account for an information stability which by any computer model should succumb to errors but does not do so in reality—this "paradox of heredity," as we called it, resists reduction upon a simple model, preformationist or epigenetic.

Physicists had previously encountered a closely similar situation. Light was a volley of particles according to some physicists, a set of waves according to others. Later, the same dualism was found to apply to electrons. In quantum mechanics this dualism is not only declared universal but is formulated in terms of rigorous mathematics and given the name "complementarity." In the biology of heredity, the only tools of description available according to our views are heterogeneous, finite classes. We now suggest that the irreducible logical complexity of biological heredity prevents us from assigning to it a specific model, in particular, the preformationist model of the "code," while eliminating all epigenetic features. This irreducible logical complexity is formally expressed in a description by heterogeneous finite classes. Instead of proposing a clever mechanism for the resolution of the paradox, we are led to think that the tools of description are limited in such a way that the paradox appears as an integral part of the description. Here, the analogy with the way physicists did overcome their own paradoxes is quite clear. In each instance of a conceptual advance in physics, the threatening paradoxes were absorbed into a modified form of description, where they no longer seem paradoxical. I here propose that the paradoxes of biology, in particular, the most widespread of them, the paradox of heredity, can be absorbed into a form of description called *holistic,* which in formal terms represents a description by heterogeneous, finite classes with its built-in limitations. Let us look at this question from still another point of view.

Niels Bohr, who was the uncontested leader among the founding fathers of quantum mechanics, discussed often in his later writings the changes in thinking that have to be made on introducing the conceptual innovations of quantum-mechanical theory. He expresses them as representing a renunciation of knowledge. Let me explain in more detail what this means. In traditional, "classical" mechanics a particle has a definite orbit, a curve in space through which it moves. As we go to quantum mechanics, this orbit is replaced by a probability distribution. The price we are paying for a new quantitative scheme of description, that of quantum mechanics, is that we are losing the knowledge of specific numerical data, in this particular case the knowledge of a well-defined orbit for the particle.

Many experiences of the physicist indicate that this way of thinking about conceptual innovation can be generalized. The renunciation of knowledge just quoted expresses a psychological condition of the scientist rather than the purely abstract replacement of one set of formal relationships by another. There are other examples from the history of physics: the requirement of the model of a spherical earth is that one renounce the knowledge of an Absolute Up and an Absolute Down, so "obvious" to the uninstructed mind and adopted without qualm by Aristotle. We see that two distinct steps are involved in a change of "model" that involves major conceptual novelty. In the first place a new formal structure is required. In the second place one needs an adaptation of the language. Bohr's renunciation of knowledge is just one aspect of this, corresponding to a somewhat negative feeling; it expresses a sort of nostalgia for old, familiar modes of thought. In biology, there are concepts which are novel in the formal theory but which in practice are old and familiar; such traits are exemplified by the paradox of heredity.

The history of conceptual innovations in physics has shown that it is all but impossible to carry through such an innovation unless a novel formal structure is available in the first place. Experience then shows that once the formal structure is agreed upon, the adaptation of the language is very much easier; in fact, I do not know of any case in the history of physics where a major adaptation of the language would have occurred in the absence of a formal scheme. In physics the formal structure has always been mathematical; in biology it is, as we have seen, *logical*. Now the aim of these pages, based on the author's previous, extensive work, was to express the necessary formal structure in terms of a set of simple propositions, as simple as we could make them. This led us to the two postulates enunciated. These postulates would be meaningless unless they could be based on the principle of finite classes, which assures us that any statement of biology is a specialization and not a generalization of the content of physics; hence, no logical contradiction between biology and physics is ever possible.

But the innovation implied by our first postulate, that of holistic selection, cannot be directly compared with the innovation necessary to go from conventional, "classical" mechanics to quantum mechanics. In the last-named case the physicist finds himself confronted with two mutually exclusive alternatives, the model of corpuscles and the model of waves. A formal scheme, consisting of the mathematics of complementarity, appears now and tempers the logical contradiction to a level acceptable to the physicist: instead of a flat logical contradiction between the picture of corpuscles and that of waves, one speaks of two aspects of reality that cannot be observed simultaneously. Experimental setups designed for the observation of waves are such that they cannot reveal the presence of particles, and vice versa.

In the case of the postulate of holistic selection, no dualism is apparent, and the same holds for the postulate of information stability. The ancient antithesis of preformation versus epigenesis expresses two systems of explanation that are logically incompatible with each other. But the innovation proposed, the selection of an admissible pattern by nature itself, does not establish a mutual exclusiveness of two alternatives. This brings us back to Bohr's "generalized complementarity" mentioned toward the end of Section IV. The selection stipulated by our postulates seems not to be related to any conflict between two mutually contradictory concepts even if it can be so formulated in terms of pure abstractions. Hence, we are better off when we conceive of the selection in a more direct sense: the selected pattern supersedes the logical division of the two models and creates in a manner of speaking a new unity, that of the organism.

Now if this is so, the question does almost at once arise of why nature goes to such great length in transmitting specific hereditary information by the use of the genetic code. Obviously, a scheme of theoretical biology which cannot make the existence of hereditary transmission by the code altogether plausible is worthless. In concluding this review, I wish to say a few words about the interpretation of the genetic code. It is usually assumed that the code transmit *the* information from one generation to the next. But what is *the* information? As we have discussed, and as is implied by the paradox of heredity, *all* information is subject to progressive degradation by errors, as time goes on. But depending on the item involved, this degradation can be more or less critical. Let me, as a very crude scheme, divide all information into "precise" components and "loose" components. Even a superficial view will indicate that much of the information required for morphology is loose, often very loose. Thus, the size of a man, or of an elephant for that matter, may change by 30%, say, without affecting the viability of the specimen appreciably. Similar remarks apply to a great many other morphological features, perhaps to their overwhelming majority.

But the function of the information embedded in the genetic code and used to reconstruct the sequence of amino acid residues in a peptide or protein molecule is altogether different. Here, only *discrete* alternatives are available. As is well known, errors in two or three of the members of the amino acid sequence will change the conformation of the resulting

A FORM OF LOGIC SUITED FOR BIOLOGY   61

enzyme sufficiently so that the enzyme loses all or most of its specific chemical activity. The enzyme is first and foremost a *tool* of the living organism. As every engineer and every toolmaker knows, the requirements to be put on a tool are much higher than those that apply to the object shaped by the tool: the tool may have to be made of hardened steel where the objects produced may consist of soft iron. We thus can gain a new insight into the nature of the information transmitted by means of the genetic code. It would be extremely "expensive" for the organism to treat the precise information in the same way the loose information of morphology is treated; it would indeed be impossible because one could not build up a specific organic structure without having *some* specific information. Evidently nature chooses the precise information as the part to be transmitted, making it easier in this way for the "epigenetic" part of the information to be regenerated by the process of holistic selection. Clearly, any such statement is a clumsy effort to express in words of the ordinary language something that cannot be so expressed.

In this contribution I have tried to utilize the experience of the physicist which tells us that a conceptual innovation cannot in practice be carried out unless a new formal scheme has in the first place been made available. But in the transition from physics to biology the innovations are logical rather than mathematical. On raising a widespread experience made in the statistical mechanics of organic compounds to a general principle, the principle of finite classes, we succeeded in reducing biological regularity to a special case of patterns compatible with physical law rather than to a logical extension of the latter. This step, as we saw, removes any contradiction between laws of physics and biological regularities. But in order to carry through such a program, we had to recognize heterogeneous, finite classes as independent formal elements of scientific description, on a par with the laws of nature so familiar to the physicist. This is a decisively novel point which defines a *program;* I can do no more here than to recommend this program to the consideration of my colleagues. It is, hopefully, a reasonably complete formal scheme, but the adaptation of language, as I called it a little while ago, corresponding to the application of the scheme to numerous concrete instances, will no doubt be a lengthy undertaking.

REFERENCES

Conant, J. B., and Blatt, A. H. (1947). "The Chemistry of Organic Compounds," 2nd ed.
Eddington, A. (1939). "The Philosophy of Physical Science." Univ. of Michigan Press, Ann Arbor.
Elsasser, W. M. (1958). "The Physical Foundation of Biology." Pergamon, Oxford.
Elsasser, W. M. (1966). "Atom and Organism." Princeton Univ. Press, Princeton, New Jersey.

Elsasser, W. M. (1975). "The Chief Abstractions of Biology." North-Holland Publ., Amsterdam.

Elsasser, W. M. (1979). *J. Soc. Biol. Struct.* **2,** 229–234.

Elsasser, W. M. (1981). *J. Theor. Biol.* **89,** 131–150.

Goodwin, B. C. (1978). *J. Soc. Biol. Struct.* **1,** 117–125.

Nordenskiöld, E. (1946). "The History of Biology." Tudor, New York.

Von Neumann, J. (1956). Probabilistic logics and the synthesis of reliable organisms from unreliable components. *In* "Automata Studies." (C. E. Shannon and J. McCarthy, eds.). Princeton Univ. Press, Princeton, New Jersey.

Williams, R. J. (1956). "Biochemical Individuality." Wiley, New York. (Reprinted by Univ. of Texas Press.)

Complexity and General Systems Theory

5. Principles of the self-organizing system

W. Ross Ashby

Originally published as Ashby, W. R. (1962). "Principles of the self-organizing system," in *Principles of Self-Organization: Transactions of the University of Illinois Symposium*, H. Von Foerster and G. W. Zopf, Jr. (eds.), Pergamon Press: London, UK, pp. 255-278. Reproduced with the kind permission of Ross Ashby's daughters, Sally Bannister, Ruth Pettit, and Jill Ashby. We would also like to thank John Ashby for his generous assistance in obtaining their permission.

The brilliant British psychiatrist, neuroscientist, and mathematician Ross Ashby was one of the pioneers in early and mid-phase cybernetics and thereby one of the leading progenitors of modern complexity theory. Not one to take either commonly used terms or popular notions for granted, Ashby probed deeply into the meaning of supposedly self-organizing systems. At the time of the following article, he had been working on a mathematical formalism of his *homeostat*, a hypothetical machine established on an axiomatic, set theoretical foundation that was supposed to offer a sufficient description of a living organism's learning and adaptive intelligence. Ashby's homeostat had a small number of essential variables serving to maintain its operation over a wide range of environmental conditions so that if the latter changed and thereby shifted the variables beyond the range where the homeostat could safely function, a new 'higher' level of the machine was activated in order to randomly reset the lower level's internal connections or organization (see Dupuy, 2000). Like the role of random mutations during evolution, if the new range set at random proved functional, the homeostat survived, otherwise it expired.

One of Ashby's goals was to repudiate that interpretation of the notion of self-organization, one commonly held to this day, which would have it that either a machine or a living organism could by itself change its own organization (or, in his phraseology, the functional mappings). For Ashby, self-organization in this sense was a bit of superfluous metaphysics since he believed not only could his formalism by itself completely delineate the homeostat's lower level organization, the adaptive novelty of his homeostat was purely the result of its upper level randomization that could reorganize the lower level and not some innate propensity for autonomous change. We offer Ashby's careful reasoning here as an enlightening guide for coming to terms with key ideas in complexity theory whose genuine significance lies less with facile bandying about and more with an intensive and extensive examination of the underlying assumptions.

Jeffrey A. Goldstein

Dupuy, J. (2000). *The Mechanization of the Mind*, Princeton: Princeton University Press.

W. ROSS ASHBY

University of Illinois

PRINCIPLES OF THE SELF-ORGANIZING SYSTEM*

Questions of principle are sometimes regarded as too unpractical to be important, but I suggest that that is certainly not the case in *our* subject. The range of phenomena that we have to deal with is so broad that, were it to be dealt with wholly at the technological or practical level, we would be defeated by the sheer quantity and complexity of it. The total range can be handled only piecemeal; among the pieces are those homomorphisms of the complex whole that we call "abstract theory" or "general principles". They alone give the bird's-eye view that enables us to move about in this vast field without losing our bearings. I propose, then, to attempt such a bird's-eye survey.

WHAT IS "ORGANIZATION"?

At the heart of our work lies the fundamental concept of "organization". What do we mean by it? As it is used in biology it is a somewhat complex concept, built up from several more primitive concepts. Because of this richness it is not readily defined, and it is interesting to notice that while March and Simon (1958) use the word "Organizations" as title for their book, they do not give a formal definition. Here I think they are right, for the word covers a multiplicity of meanings. I think that in future we shall hear the *word* less frequently, though the *operations* to which it corresponds, in the world of computers and brain-like mechanisms, will become of increasing daily importance.

The hard core of the concept is, in my opinion, that of "conditionality". As soon as the relation between two entities *A* and *B*

---

* The work on which this paper is based was supported by ONR Contract N 049–149.

255

becomes conditional on C's value or state then a necessary component of "organization" is present. Thus *the theory of organization
is partly co-extensive with the theory of functions of more than one
variable*.

We can get another angle on the question by asking "what is
its converse?" The converse of "conditional on" is "not conditional on", so the converse of "organization" must therefore be,
as the mathematical theory shows as clearly, the concept of
"reducibility". (It is also called "separability".) This occurs, in
mathematical forms, when what looks like a function of several
variables (perhaps very many) proves on closer examination to
have parts whose actions are *not* conditional on the values of the
other parts. It occurs in mechanical forms, in hardware, when
what looks like one machine proves to be composed of two (or
more) sub-machines, each of which is acting independently of
the others.

Questions of "conditionality", and of its converse "reducibility",
can, of course, be treated by a number of mathematical and logical
methods. I shall say something of such methods later. Here,
however, I would like to express the opinion that the method of
Uncertainty Analysis, introduced by Garner and McGill (1956),
gives us a method for the treatment of conditionality that is not
only completely rigorous but is also of extreme generality. Its great
generality and suitability for application to complex behavior,
lies in the fact that it is applicable to any arbitrarily defined set of
states. Its application requires neither linearity, nor continuity,
nor a metric, nor even an ordering relation. By this calculus, the
degree of conditionality can be measured, and analyzed, and
apportioned to factors and interactions in a manner exactly parallel
to Fisher's method of the analysis of variance; yet it requires no
metric in the variables, only the frequencies with which the various
combinations of states occur. It seems to me that, just as Fisher's
conception of the analysis of variance threw a flood of light on to
the complex relations that may exist between variations on a
metric, so McGill and Garner's conception of uncertainty analysis
may give us an altogether better understanding of how to treat
complexities of relation when the variables are non-metric. In
psychology and biology such variables occur with great commonness; doubtless they will also occur commonly in the brain-like

processes developing in computers. I look forward to the time when the methods of McGill and Garner will become the accepted language in which such matters are to be thought about and treated quantitatively.

The treatment of "conditionality" (whether by functions of many variables, by correlation analysis, by uncertainty analysis, or by other ways) makes us realize that the essential idea is that there is first a product space—that of the *possibilities*—within which some sub-set of points indicates the actualities. This way of looking at "conditionality" makes us realize that it is related to that of "communication"; and it is, of course, quite plausible that we should define parts as being "organized" when "communication" (in some generalized sense) occurs between them. (Again the natural converse is that of independence, which represents non-communication.)

Now "communication" from *A* to *B* necessarily implies some constraint, some correlation between what happens at *A* and what at *B*. If, for given event at *A*, all possible events may occur at *B*, then there is no communication from *A* to *B* and no constraint over the possible (*A*, *B*)-couples that can occur. Thus the presence of "organization" between variables is equivalent to the existence of a *constraint* in the product-space of the possibilities. I stress this point because while, in the past, biologists have tended to think of organization as something extra, something *added* to the elementary variables, the modern theory, based on the logic of communication, regards organization as a restriction or constraint. The two points of view are thus diametrically opposed; there is no question of either being exclusively right, for each can be appropriate in its context. But with this opposition in existence we must clearly go carefully, especially when we discuss with others, lest we should fall into complete confusion.

This excursion may seem somewhat complex but it is, I am sure, advisable, for we have to recognize that the discussion of organization theory has a peculiarity not found in the more objective sciences of physics and chemistry. The peculiarity comes in with the product space that I have just referred to. Whence comes this product space? Its chief peculiarity is that *it contains more than actually exists in the real physical world*, for it is the latter that gives us the actual, constrained *subset*.

18

W. ROSS ASHBY

The real world gives the subset of what *is*; the product space represents the uncertainty of the *observer*. The product space may therefore change if the observer changes; and two observers may legitimately use different product spaces within which to record the same subset of actual events in some actual thing. The "constraint" is thus a *relation* between observer and thing; the properties of any particular constraint will depend on both the real thing and on *the observer*. It follows that a substantial part of the theory of organization will be concerned with *properties that are not intrinsic to the thing but are relational between observer and thing*. We shall see some striking examples of this fact later.

WHOLE AND PARTS

"If conditionality" is an essential component in the concept of organization, so also is the assumption that we are speaking of a whole composed of parts. This assumption is worth a moment's scrutiny, for research is developing a theory of dynamics that does *not* observe parts and their interactions, but treats the system as an unanalysed whole (Ashby, 1958, a). In physics, of course, we usually start the description of a system by saying "Let the variables be $x_1, x_2,..., x_n$" and thus start by treating the whole as made of n functional parts. The other method, however, deals with unanalysed states, $S_1, S_2,...$ of the whole, without explicit mention of any parts that may be contributing to these states. The dynamics of such a system can then be defined and handled mathematically; I have shown elsewhere (Ashby, 1960, a) how such an approach can be useful. What I wish to point out here is that we can have a sophisticated *dynamics*, of a whole as complex and cross-connected as you please, that makes no reference to any parts and that therefore does *not* use the concept of organization. Thus the concepts of dynamics and of organization are essentially independent, in that all four combinations, of their presence and absence, are possible.

This fact exemplifies what I said, that "organization" is partly in the eye of the beholder. Two observers studying the same real material system, a hive of bees say, may find that one of them, thinking of the hive as an interaction of fifty thousand bee-parts, finds the bees "organized", while the other, observing whole states such as activity, dormancy, swarming, etc., may see *no* organization, only trajectories of these (unanalysed) states.

Another example of the independence of "organization" and "dynamics" is given by the fact that whether or not a real system is organized or reducible depends partly on the point of view taken by the observer. It is well known, for instance, that an organized (i.e. interacting) linear system of *n* parts, such as a network of pendulums and springs, can be seen from another point of view (that of the so-called "normal" coordinates) in which all the (newly identified) parts are completely separate, so that the whole is reducible. There is therefore nothing perverse about my insistence on the relativity of organization, for advantage of the fact is routinely taken in the study of quite ordinary dynamic systems.

Finally, in order to emphasize how dependent is the organization seen in a system on the observer who sees it, I will state the proposition that: given a whole with arbitrarily given behavior, a great variety of arbitrary "parts" can be seen in it; for all that is necessary, when the arbitrary part is proposed, is that we assume the given part to be coupled to another suitably related part, so that the two together form a whole isomorphic with the whole that was given. For instance, suppose the given whole, W of 10 states, behaves in accordance with the transformation:

$$W \downarrow \begin{array}{c} p\ q\ r\ s\ t\ u\ v\ w\ x\ y \\ q\ r\ s\ q\ s\ t\ t\ x\ y\ y \end{array}$$

Its kinematic graph is

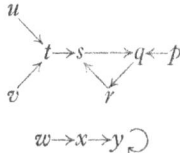

$$u \searrow$$
$$\nearrow\ t \rightarrow s \dashrightarrow q \leftarrow p$$
$$v \qquad\quad \searrow$$
$$\qquad\qquad r$$

$$w \rightarrow x \rightarrow y \circlearrowright$$

and suppose we wish to "see" it as containing the part P, with internal states E and input states A:

$$\left.\begin{array}{cc|cc}
 & & \multicolumn{2}{c}{E} \\
 & \downarrow & 1 & 2 \\
\hline
 & 1 & 2 & 1 \\
A & 2 & 1 & 1 \\
\end{array}\right\} P$$

W. ROSS ASHBY

With a little ingenuity we find that if part P is coupled to part Q (with states (F, G) and input B) with transformation Q:

$$(F, G)$$

$\downarrow$		1, 1	1, 2	1, 3	2, 1	2, 2	2, 3	
	1	2, 1	1, 2	1, 2	2, 1	1, 2	1, 2	$\Big\} Q$
B	2	·	2, 3	·	2, 1	2, 2	2, 2	

by putting $A = F$ and $B = E$, then the new whole W' has transformation

W': $\quad \downarrow \quad$
$\begin{array}{llll} 1,1,1 & 1,1,2 & 1,1,3 & 1,2,1, \text{ etc.} \\ 2,2,1 & 2,1,2 & 2,1,2 & 1,2,1, \text{ etc.} \end{array}$

which is *isomorphic with* W under the one–one correspondence

$\downarrow$
$\begin{array}{llll} 1,1,1 & 1,1,2 & 1,1,3 & 1,2,1, \text{ etc.} \\ w & s & p & y \quad , \text{ etc.} \end{array}$

Thus, subject only to certain requirements (e.g. that equilibria map into equilibria) *any dynamic system can be made to display a variety of arbitrarily assigned "parts"*, simply by a change in the *observer's* view point.

MACHINES IN GENERAL

I have just used a way of representing two "parts", "coupled" to form a "whole", that anticipates the question: what do we mean by a "machine" in general?

Here we are obviously encroaching on what has been called "general system theory", but this last discipline always seemed to me to be uncertain whether it was dealing with *physical* systems, and therefore tied to whatever the real world provides, or with mathematical systems, in which the sole demand is that the work shall be free from internal contradictions. It is, I think, one of the substantial advances of the last decade that we have at last identified the *essentials* of the "machine in general".

Before the essentials could be seen, we had to realize that two factors must be *excluded as irrelevant*. The first is "materiality"—the idea that a machine must be made of actual matter, of the hundred or so existent elements. This is wrong, for examples can readily be given (e.g. Ashby, 1958, a) showing that what is essential is whether the system, of angels and ectoplasm if you please, *behaves* in a law-abiding and machine-like way. Also to be excluded as irrelevant is any reference to energy, for any calculating machine shows that what matters is the *regularity* of the behavior—whether energy is gained or lost, or even created, is simply irrelevant.

The fundamental concept of "machine" proves to have a form that was formulated at least a century ago, but this concept has not, so far as I am aware, ever been used and exploited vigorously. A "machine" is that which behaves in a machine-like way, namely, that its internal state, and the state of its surroundings, defines uniquely the next state it will go to.

This definition, formally proposed fifteen years ago (Ashby, 1945) has withstood the passage of time and is now becoming generally accepted (e.g. Jeffrey, 1959). It appears in many forms. When the variables are continuous it corresponds to the description of a dynamic system by giving a set of ordinary differential equations with time as the independent variable. The *fundamental* nature of such a representation (as contrasted with a merely convenient one) has been recognized by many earlier workers such as Poincaré, Lotka (1925), and von Bertalanffy (1950 and earlier).

Such a representation by differential equations is, however, too restricted for the needs of a science that includes biological systems and calculating machines, in which discontinuity is ubiquitous. So arises the modern definition, able to include both the continuous and the discontinuous and even the discrete, without the slightest loss of rigor. The "machine with input" (Ashby, 1958, a) or the "finite automaton" (Jeffrey, 1959) is today defined by a set S of internal states, a set I of input or surrounding states, and a mapping, f say, of the product set $I \times S$ into S. Here, in my opinion, we have the very essence of the "machine"; all known types of machine are to be found here; and all interesting deviations from the concept are to be found by the corresponding deviation from the definition.

We are now in a position to say without ambiguity or evasion what we mean by a machine's "organization". First we specify which system we are talking about by specifying its states S and its conditions I. If S is a product set, so that $S = \Pi_i T_i$ say, then the parts i are each specified by its set of states T_i. *The "organization" between these parts is then specified by the mapping f.* Change f and the organization changes. In other words, the possible organizations between the parts can be set into one–one correspondence with the set of possible mappings of $I \times S$ into S. Thus "organization" and "mapping" are two ways of looking at the same thing—the organization being noticed by the observer of the actual system, and the mapping being recorded by the person who represents the behavior in mathematical or other symbolism.

"GOOD" ORGANIZATION

At this point some of you, especially the biologists, may be feeling uneasy; for this definition of organization makes no reference to any *usefulness* of the organization. It demands only that there be conditionality between the parts and regularity in behavior. In this I believe the definition to be right, for the question whether a given organization is "good" or "bad" is quite independent of the prior test of whether it is or is not an organization.

I feel inclined to stress this point, for here the engineers and the biologists are likely to think along widely differing lines. The engineer, having put together some electronic hardware and having found the assembled network to be roaring with parasitic oscillations, is quite accustomed to the idea of a "bad" organization; and he knows that the "good" organization has to be searched for. The biologist, however, studies mostly animal species that have survived the long process of natural selection; so almost all the organizations he sees have already been selected to be good ones, and he is apt to think of "organizations" as *necessarily* good. This point of view may often be true in the biological world but it is most emphatically not true in the world in which we people here are working. We *must* accept that

(1) most organizations are bad ones;

(2) the good ones have to be sought for; and

(3) what is meant by "good" must be clearly defined, explicitly if necessary, *in every case*.

What then is meant by "good", in our context of brain-like mechanisms and computers? We must proceed cautiously, for the word suggests some evaluation whose origin has not yet been considered.

In some cases the distinction between the "good" organization and the "bad" is obvious, in the sense that as everyone in these cases would tend to use the same criterion, it would not need explicit mention. The brain of a living organism, for instance, is usually judged as having a "good" organization if the organization (whether inborn or learned) acts so as to further the organism's survival. This consideration readily generalizes to all those cases in which the organization (whether of a cat or an automatic pilot or an oil refinery) is judged "good" if and only if it acts so as to keep an assigned set of variables, the "essential" variables, within assigned limits. Here are all the mechanisms for homeostasis, both in the original sense of Cannon and in the generalized sense. From this criterion comes the related one that an organization is "good" if it makes the system stable around an assigned equilibrium. Sommerhoff (1950) in particular has given a wealth of examples, drawn from a great range of biological and mechanical phenomena, showing how in all cases the idea of a "good organization" has as its essence the idea of a number of parts so interacting as to achieve some given "focal condition". I would like to say here that I do not consider that Sommerhoff's contribution to our subject has yet been adequately recognized. His identification of *exactly* what is meant by coordination and integration is, in my opinion, on a par with Cauchy's identification of exactly what was meant by convergence. Cauchy's discovery was a real discovery, and was an enormous help to later workers by providing them with a concept, rigorously defined, that could be used again and again, in a vast range of contexts, and always with exactly the same meaning. Sommerhoff's discovery of how to represent *exactly* what is meant by coordination and integration and good organization will, I am sure, eventually play a similarly fundamental part in our work.

His work illustrates, and emphasizes, what I want to say here— *there is no such thing as "good organization" in any absolute sense.* Always it is relative; and an organization that is good in one context or under one criterion may be bad under another.

Sometimes this statement is so obvious as to arouse no opposition. If we have half a dozen lenses, for instance, that can be

264   W. ROSS ASHBY

assembled this way to make a telescope or that way to make a microscope, the goodness of an assembly obviously depends on whether one wants to look at the moon or a cheese mite.

But the subject is more contentious than that! The thesis implies that there is no such thing as a brain (natural or artificial) that is good in any absolute sense—it all depends on the circumstances and on what is wanted. Every faculty that a brain can show is "good" only conditionally, for there exists at least one environment against which the brain is handicapped by the possession of this faculty. Sommerhoff's formulation enables us to show this at once: whatever the faculty or organization achieves, let that be *not* in the "focal conditions".

We know, of course, lots of examples where the thesis is true in a somewhat trivial way. Curiosity tends to be good, but many an antelope has lost its life by stopping· to see what the hunter's hat is. Whether the organization of the antelope's brain should be of the type that does, or does not, lead to temporary immobility clearly depends on whether hunters with rifles are or are not plentiful in its world.

From a different angle we can notice Pribram's results (1957), who found that brain-operated monkeys scored higher in a certain test than the normals. (The operated were plodding and patient while the normals were restless and distractible.) Be that as it may, one cannot say which brain (normal or operated) had the "good" organization until one has decided which sort of temperament is wanted.

Do you still find this non-contentious? Then I am prepared to assert that there is not a single mental faculty ascribed to Man that is good in the absolute sense. If any particular faculty is *usually* good, this is solely because our terrestrial environment is so lacking in variety that its usual form makes that faculty usually good. But change the environment, go to really different conditions, and possession of that faculty may be harmful. And "bad", by implication, is the brain organization that produces it.

I believe that there is not a single faculty or property of the brain, usually regarded as desirable, that does not become *un*desirable in some type of environment. Here are some examples in illustration.

The first is Memory. Is it not good that a brain should have memory? Not at all, I reply—only when the environment is of a type in which the future often *copies* the past; should the future often be the *inverse* of the past, memory is actually disadvantageous. A well known example is given when the sewer rat faces the environmental system known as "pre-baiting". The naïve rat is very suspicious, and takes strange food only in small quantities. If, however, wholesome food appears at some place for three days in succession, the sewer rat will learn, and on the fourth day will eat to repletion, and die. The rat without memory, however, is as suspicious on the fourth day as on the first, and lives. Thus, in *this* environment, memory is positively disadvantageous. Prolonged contact with this environment will lead, other things being equal, to evolution in the direction of diminished memory-capacity.

As a second example, consider organization itself in the sense of connectedness. Is it not good that a brain should have its parts in rich functional connection? I say, No—not *in general*; only when the environment is itself richly connected. When the environment's parts are *not* richly connected (when it is highly reducible, in other words), adaptation will go on faster if the brain is also highly reducible, i.e. if its connectivity is small (Ashby, 1960, d). Thus the *degree* of organization can be too high as well as too low; the degree we humans possess is probably adjusted to be somewhere near the optimum for the usual terrestrial environment. It does not in any way follow that this degree will be optimal or good if the brain is a mechanical one, working against some grossly non-terrestrial environment—one existing only inside a big computer, say.

As another example, what of the "organization" that the biologist always points to with pride—the development in evolution of specialized organs such as brain, intestines, heart and blood vessels. Is not this good? Good or not, it is certainly a specialization made possible only because the earth has an atmosphere; without it, we would be incessantly bombarded by tiny meteorites, any one of which, passing through our chest, might strike a large blood vessel and kill us. Under such conditions a better form for survival would be the slime mould, which specializes in being able to flow through a tangle of twigs without loss of function. Thus the development of organs is not good unconditionally, but is a specialization to a world free from flying particles.

W. ROSS ASHBY

After these actual instances, we can return to theory. It is here that Sommerhoff's formulation gives such helpful clarification. He shows that in all cases there must be given, and specified, first a *set of disturbances* (values of his "coenetic variable") and secondly a goal (his "focal condition"); the disturbances threaten to drive the outcome outside the focal condition. The "good" organization is then of the nature of a *relation* between the set of disturbances and the goal. Change the set of disturbances, and the organization, without itself changing, is evaluated "bad" instead of "good". As I said, there is no property of an organization that is good in any absolute sense; all are relative to some given environment, or to some given set of threats and disturbances, or to some given set of problems.

SELF-ORGANIZING SYSTEMS

I hope I have not wearied you by belaboring this relativity too much, but it is fundamental, and is only too readily forgotten when one comes to deal with organizations that are either biological in origin or are in imitation of such systems. With this in mind, we can now start to consider the so-called "self-organizing" system. We must proceed with some caution here if we are not to land in confusion, for the adjective is, if used loosely, ambiguous, and, if used precisely, self-contradictory.

To say a system is "self-organizing" leaves open two quite different meanings.

There is a first meaning that is simple and unobjectionable. This refers to the system that starts with its parts separate (so that the behavior of each is independent of the others' states) and whose parts then act so that they change towards forming connections of some type. Such a system is "self-organizing" in the sense that it changes from "parts separated" to "parts joined". An example is the embryo nervous system, which starts with cells having little or no effect on one another, and changes, by the growth of dendrites and formation of synapses, to one in which each part's behavior is very much affected by the other parts. Another example is Pask's system of electrolytic centers, in which the growth of a filament from one electrode is at first little affected by growths at the other electrodes; then the growths become more and more affected by one another as filaments approach the other electrodes. In general such systems can be more simply characterized as "self-*connecting*", for the change from independence between the parts to conditionality can always be seen as some form of "connection", even if it is as purely functional as that from a radio transmitter to a receiver.

Here, then, is a perfectly straightforward form of self-organizing system; but I must emphasize that there can be no assumption at this point that the organization developed will be a good one. If we wish it to be a "good" one, we must first provide a criterion for distinguishing between the bad and the good, and then we must ensure that the appropriate selection is made.

We are here approaching the second meaning of "self-organizing" (Ashby, 1947). "Organizing" may have the first meaning, just discussed, of "changing from unorganized to organized". But it may also mean "changing from a bad organization to a good one", and this is the case I wish to discuss now, and more fully. This is the case of peculiar interest to *us*, for this is the case of the system that changes itself from a bad way of behaving to a good. A well known example is the child that starts with a brain organization that makes it fire-seeking; then a change occurs, and a new brain organization appears that makes the child fire-avoiding. Another example would occur if an automatic pilot and a plane were so coupled, by mistake, that positive feedback made the whole error-aggravating rather than error-correcting. Here the organization is bad. The system would be "self-organizing" if a change were *automatically* made to the feedback, changing it from positive to negative; then the whole would have changed from a bad organization to a good. Clearly, *this* type of "self-organization" is of peculiar interest to us. What is implied by it?

Before the question is answered we must notice, if we are not to be in perpetual danger of confusion, that *no machine can be self-organizing in this sense*. The reasoning is simple. Define the set S of states so as to specify which machine we are talking about. The "organization" must then, as I said above, be identified with f, the mapping of S into S that the basic drive of the machine (whatever force it may be) imposes. Now the logical relation here is that f determines the changes of S:—f is *defined* as the set of couples (s_i, s_j) such that the internal drive of the system will force state s_i to change to s_j. To allow f to be a function of the state is to make nonsense of the whole concept.

Since the argument is fundamental in the theory of self-organizing systems, I may help explanation by a parallel example. Newton's law of gravitation says that $F = M_1 M_2 / d^2$, in particular, that the force varies inversely as the distance to power 2. To power 3 would be a different law. But suppose it were suggested that, not the force F but the *law* changed with the distance, so that the power was not 2 but some function of the distance, $\phi(d)$. This suggestion is illogical; for we now have that $F = M_1 M_2 / d^{\phi(d)}$, and this represents not a law that varies with the distance but *one* law covering all distances; that is, were this the case we would *re-define* the law. Analogously, were f in the machine to be some function of the state S, we would have to re-define our machine. Let me be quite explicit with an example. Suppose S had three states: a, b, c. If f depended on S there would be three f's: f_a, f_b, f_c say. Then if they are

$\downarrow$	a	b	c
f_a	**b**	a	b
f_b	c	**a**	a
f_c	b	b	**a**

then the transform of a must be under f_a, and is therefore b, so the whole set of f's would amount to the *single* transformation:

$$\downarrow \quad \begin{matrix} a & b & c \\ b & a & a \end{matrix}$$

It is clearly illogical to talk of f as being a function of S, for such talk would refer to operations, such as $f_a(b)$, which cannot in fact occur.

If, then, no machine can properly be said to be self-organizing, how do we regard, say, the Homeostat, that rearranges its own wiring; or the computer that writes out its own program?

The new logic of mechanism enables us to treat the question rigorously. We start with the set S of states, and assume that f changes, to g say. So we really have a *variable*, $\alpha(t)$ say, a function of the time that had at first the value f and later the value g. This change, as we have just seen, cannot be ascribed to any cause in the set S; so it must have come from some outside agent, acting on the system S as input. If the system is to be in some sense "*self-*organizing", the "self" must be enlarged to include this variable α, and, to keep the whole bounded, the cause of α's change must be in S (or α).

Thus the appearance of being "self-organizing" can be given only by the machine S being coupled to another machine (of one part):

$$\boxed{S} \xrightarrow{\quad} \boxed{\alpha}$$
$$\xleftarrow{\quad}$$

Then the part S can be "self-organizing" within the whole $S + \alpha$.

Only in this partial and strictly qualified sense can we understand that a system is "*self*-organizing" without being self-contradictory.

Since no system can correctly be said to be self-organizing, and since use of the phrase "self-organizing" tends to perpetuate a fundamentally confused and inconsistent way of looking at the subject, the phrase is probably better allowed to die out.

THE SPONTANEOUS GENERATION OF ORGANIZATION

When I say that no system can properly be said to be self-organizing, the listener may not be satisfied. What, he may ask, of those changes that occurred a billion years ago, that led lots of carbon atoms, scattered in little molecules of carbon dioxide, methane, carbonate, etc., to get together until they formed proteins, and then went on to form those large active lumps that today we call "animals"? Was not this process, on an isolated planet, one of "self-organization"? And if it occurred on a planetary surface can it not be made to occur in a computer? I am, of course, now discussing the origin of life. Has modern system theory anything to say on this topic?

It has a great deal to say, and some of it flatly contradictory to what has been said ever since the idea of evolution was first considered. In the past, when a writer discussed the topic, he usually assumed that the generation of life was rare and peculiar,

270   W. ROSS ASHBY

and he then tried to display some way that would enable this rare and peculiar event to occur. So he tried to display that there is *some* route from, say, carbon dioxide to the amino acid, and thence to the protein, and so, through natural selection and evolution, to intelligent beings. I say that this looking for special conditions is quite wrong. The truth is the opposite—*every* dynamic system generates its own form of intelligent life, is self-organizing in this sense. (I will demonstrate the fact in a moment.) Why we have failed to recognize this fact is that until recently we have had no experience of systems of medium complexity; either they have been like the watch and the pendulum, and we have found their properties few and trivial, or they have been like the dog and the human being, and we have found their properties so rich and remarkable that we have thought them supernatural. Only in the last few years has the general-purpose computer given us a system rich enough to be interesting yet still simple enough to be understandable. With this machine as tutor we can now begin to think about systems that are simple enough to be comprehensible in detail yet also rich enough to be suggestive. With their aid we can see the truth of the statement that *every isolated determinate dynamic system obeying unchanging laws will develop "organisms" that are adapted to their "environments"*.

The argument is simple enough in principle. We start with the fact that systems in general go to equilibrium. Now most of a system's states are non-equilibrial (if we exclude the extreme case of the system in neutral equilibrium). So in going from *any* state to one of the equilibria, the system is going from a larger number of states to a smaller. In this way it is performing a selection, in the purely objective sense that it rejects some states, by leaving them, and retains some other state, by sticking to it. Thus, as every determinate system goes to equilibrium, so does it select. We have heard *ad nauseam* the dictum that a machine cannot select; the truth is just the opposite: every machine, as it goes to equilibrium, performs the corresponding act of selection.

Now, equilibrium in simple systems is usually trivial and uninteresting; it is the pendulum hanging vertically; it is the watch with its main-spring run down; the cube resting flat on one face. Today, however, we know that when the system is more complex and dynamic, equilibrium, and the stability around it, can be much more interesting. Here we have the automatic pilot success-fully combating an eddy; the person redistributing his blood flow after a severe haemorrhage; the business firm restocking after a sudden increase in consumption; the economic system restoring a distribution of supplies after a sudden destruction of a food crop; and it is a man successfully getting at least one meal a day during a lifetime of hardship and unemployment.

What makes the change, from trivial to interesting, is simply the *scale* of the events. "Going to equilibrium" *is* trivial in the simple pendulum, for the equilibrium is no more than a single point. But when the system is more complex; when, say, a country's economy goes back from wartime to normal methods then the stable region is vast, and much interesting activity can occur within it. The computer is heaven-sent in this context, for it enables us to bridge the enormous conceptual gap from the simple and understandable to the complex and interesting. Thus we can gain a considerable insight into the so-called spontaneous genera-tion of life by just seeing how a somewhat simpler version will appear in a computer.

COMPETITION

Here is an example of a simpler version. The competition between species is often treated as if it were essentially biological; it is in fact an expression of a process of far greater generality. Suppose we have a computer, for instance, whose stores are filled at random with the digits 0 to 9. Suppose its dynamic law is that the digits are continuously being multiplied in pairs, and the right-hand digit of the product going to replace the first digit taken. Start the machine, and let it "evolve"; what will happen? Now under the laws of this particular world, even times even gives even, and odd times odd gives odd. But even times odd gives even; so after a mixed encounter *the even has the better chance of survival.* So as this system evolves, we shall see the evens favored in the struggle, steadily replacing the odds in the stores and eventually exterminating them.

But the evens are not homogeneous, and among them the zeros are best suited to survive in this particular world; and, as we watch, we shall see the zeros exterminating their fellow-evens, until eventually they inherit this particular earth.

What we have here is an example of a thesis of extreme generality. From one point of view we have simply a well defined operator (the multiplication and replacement law) which drives on towards equilibrium. In doing so it *automatically* selects those operands that are *specially resistant* to its change-making tendency (for the zeros are uniquely resistant to change by multiplication). This process, of progression towards the specially resistant form, is of extreme generality, demanding only that the operator (or the physical laws of any physical system) be determinate and unchanging. This is the general or abstract point of view. The biologist sees a special case of it when he observes the march of evolution, survival of the fittest, and the inevitable emergence of the highest biological functions and intelligence. Thus, when we ask: What was necessary that life and intelligence should appear? the answer is not carbon, or amino acids or any other special feature but only that the dynamic laws of the process should be *unchanging*, i.e. that the system should be *isolated. In any isolated system, life and intelligence inevitably develop* (they may, in degenerate cases, develop to only zero degree).

So the answer to the question: How can we generate intelligence synthetically? is as follows. Take a dynamic system whose laws are unchanging and single-valued, and whose size is so large that after it has gone to an equilibrium that involves only a small fraction of its total states, this small fraction is still large enough to allow room for a good deal of change and behavior. Let it go on for a long enough time to get to such an equilibrium. Then examine the equilibrium in detail. You will find that the states or forms now in being are peculiarly able to survive against the changes induced by the laws. Split the equilibrium in two, call one part "organism" and the other part "environment": you will find that this "organism" is peculiarly able to survive against the disturbances from this "environment". The *degree* of adaptation and complexity that this organism can develop is bounded only by the size of the whole dynamic system and by the time over which it is allowed to progress towards equilibrium. Thus, as I said, every isolated determinate dynamic system will develop organisms that are adapted to their environments. There is thus no difficulty in principle, in developing synthetic organisms as complex or as intelligent as we please.

In *this* sense, then, *every* machine can be thought of as "self-organizing", for it will develop, to such degree as its size and complexity allow, some functional structure homologous with an "adapted organism". But does this give us what we at this Conference are looking for? Only partly; for nothing said so far has any implication about the organization being good or bad; the criterion that would make the distinction has not yet been introduced. It is true, of course, that the developed organism, being stable, will have its own essential variables, and it will show its stability by vigorous reactions that tend to preserve its own existence. To *itself*, its own organization will *always*, by definition, be good. The wasp finds the stinging reflex a good thing, and the leech finds the blood-sucking reflex a good thing. But these criteria come *after* the organization for survival; having seen *what* survives we then see what is "good" for that form. What emerges depends simply on what are the system's laws and from what state it started; there is no implication that the organization developed will be "good" in any absolute sense, or according to the criterion of any outside body such as ourselves.

To summarize briefly: there is no difficulty, in principle, in developing *synthetic organisms as complex, and as intelligent as we please*. But we must notice two fundamental qualifications; first, their intelligence will be an adaptation to, and a specialization towards, their particular environment, with no implication of validity for any other environment such as ours; and secondly, their intelligence will be directed towards keeping their own essential variables within limits. They will be fundamentally selfish. So we now have to ask: In view of these qualifications, can we yet turn these processes to our advantage?

REQUISITE VARIETY

In this matter I do not think enough attention has yet been paid to Shannon's Tenth Theorem (1949) or to the simpler "law of requisite variety" in which I have expressed the same basic idea (Ashby, 1958, a). Shannon's theorem says that if a correction-channel has capacity H, then equivocation of amount H can be

19

274   W. ROSS ASHBY

removed, *but no more*. Shannon stated his theorem in the context
of telephone or similar communication, but the formulation is
just as true of a biological regulatory channel trying to exert some
sort of corrective control. He thought of the case with a lot of
message and a little error; the biologist faces the case where the
"message" is small but the disturbing errors are many and large.
The theorem can then be applied to the brain (or any other
regulatory and selective device), when it says that the amount of
regulatory or selective action that the brain can achieve is absolutely bounded by its capacity as a channel (Ashby, 1958, b).
Another way of expressing the same idea is to say that any quantity
K of appropriate selection demands the transmission or processing
of quantity K of information (Ashby, 1960, b.) *There is no getting
of selection for nothing.*

I think that here we have a principle that we shall hear much
of in the future, for it dominates all work with complex systems.
It enters the subject somewhat as the law of conservation of
energy enters power engineering. When that law first came in,
about a hundred years ago, many engineers thought of it as a
disappointment, for it stopped all hopes of perpetual motion.
Nevertheless, it did in fact lead to the great practical engineering
triumphs of the nineteenth century, because it made power
engineering more realistic.

I suggest that when the full implications of Shannon's Tenth
Theorem are grasped we shall be, first sobered, and then helped,
for we shall then be able to focus our activities on the problems
that are properly realistic, and actually solvable.

THE FUTURE

Here I have completed this bird's-eye survey of the principles
that govern the self-organizing system. I hope I have given justification for my belief that these principles, based on the logic of
mechanism and on information theory, are now essentially
complete, in the sense that there is now no area that is grossly
mysterious.

Before I end, however, I would like to indicate, very briefly,
the directions in which future research seems to me to be most
likely to be profitable.

One direction in which I believe a great deal to be readily discoverable, is in the discovery of new types of dynamic process. Most of the machine-processes that we know today are very specialized, depending on exactly what parts are used and how they are joined together. But there are systems of more net-like construction in which what happens can only be treated statistically. There are processes here like, for instance, the spread of epidemics, the fluctuations of animal populations over a territory, the spread of wave-like phenomena over a nerve-net. These processes are, in themselves, neither good nor bad, but they exist, with all their curious properties, and doubtless the brain will use them should they be of advantage. What I want to emphasize here is that they often show very surprising and peculiar properties; such as the tendency, in epidemics, for the outbreaks to occur in waves. Such peculiar new properties may be just what some machine designer wants, and that he might otherwise not know how to achieve.

The study of such systems must be essentially statistical, but this does not mean that each system must be individually stochastic. On the contrary, it has recently been shown (Ashby, 1960, c) that no system can have greater efficiency than the determinate when acting as a regulator; so, as regulation is the one function that counts biologically, we can expect that natural selection will have made the brain as determinate as possible. It follows that we can confine our interest to the lesser range in which the sample space is over a set of mechanisms each of which is individually determinate.

As a particular case, a type of system that deserves much more thorough investigation is the large system that is built of parts that have many states of equilibrium. Such systems are extremely common in the terrestrial world; they exist all around us, and in fact, intelligence as we know it would be almost impossible otherwise (Ashby, 1960, d). This is another way of referring to the system whose variables behave largely as part-functions. I have shown elsewhere (Ashby, 1960, a) that such systems tend to show habituation (extinction) and to be able to adapt progressively (Ashby, 1960, d). There is reason to believe that some of the well-known but obscure biological phenomena such as conditioning, association, and Jennings' (1906) law of the resolution of physiological states may be more or less simple and direct expressions

W. ROSS ASHBY

of the multiplicity of equilibrial states. At the moment I am investigating the possibility that the transfer of "structure", such as that of three-dimensional space, into a dynamic system— the sort of learning that Piaget has specially considered—may be an *automatic* process when the input comes to a system with many equilibria. Be that as it may, there can be little doubt that the study of such systems is likely to reveal a variety of new dynamic processes, giving us dynamic resources not at present available.

A particular type of system with many equilibria is the system whose parts have a high "threshold"—those that tend to stay at some "basic" state unless some function of the input exceeds some value. The general properties of such systems is still largely unknown, although Beurle (1956) has made a most interesting start. They deserve extensive investigation; for, with their basic tendency to develop avalanche-like waves of activity, their dynamic properties are likely to prove exciting and even dramatic. The fact that the mammalian brain uses the property extensively suggests that it may have some peculiar, and useful, property not readily obtainable in any other way.

Reference to the system with many equilibria brings me to the second line of investigation that seems to me to be in the highest degree promising—I refer to the discovery of *the living organism's memory store*: the identification of its physical nature.

At the moment, our knowledge of the living brain is grossly out of balance. With regard to what happens from one millisecond to the next we know a great deal, and many laboratories are working to add yet more detail. But when we ask what happens in the brain from one hour to the next, or from one year to the next, practically nothing is known. Yet it is these longer-term changes that are the really significant ones in human behavior.

It seems to me, therefore, that if there is one thing that is crying out to be investigated it is the physical basis of the brain's memorystores. There was a time when "memory" was a very vague and metaphysical subject; but those days are gone. "Memory", as a *constraint* holding over events of the past and the present, and a *relation* between them, is today firmly grasped by the logic of mechanism. We know exactly what we mean by it behavioristically and operationally. What we need now is the provision of adequate resources for its investigation. Surely the time has come for the world to be able to find resources for *one* team to go into the matter?

SUMMARY

Today, the principles of the self-organizing system are known with some completeness, in the sense that no major part of the subject is wholly mysterious.

We have a secure base. Today we know *exactly* what we mean by "machine", by "organization", by "integration", and by "self-organization". We understand these concepts as thoroughly and as rigorously as the mathematician understands "continuity" or "convergence".

In these terms we can see today that the artificial generation of dynamic systems with "life" and "intelligence" is not merely simple—it is unavoidable if only the basic requirements are met. These are not carbon, water, or any other material entities but the persistence, over a long time, of the action of any operator that is both unchanging and single-valued. *Every* such operator forces the development of its own form of life and intelligence.

But will the forms developed be of use to *us*? Here the situation is dominated by the basic law of requisite variety (and Shannon's Tenth Theorem), which says that the achieving of appropriate selection (to a degree better than chance) is absolutely dependent on the processing of at least that quantity of information. Future work must respect this law, or be marked as futile even before it has started.

Finally, I commend as a program for research, the *identification of the physical basis of the brain's memory stores*. Our knowledge of the brain's functioning is today grossly out of balance. A vast amount is known about how the brain goes from state to state at about millisecond intervals; but when we consider our knowledge of the basis of the important long-term changes we find it to amount, practically, to nothing. I suggest it is time that we made some definite attempt to attack this problem. Surely it is time that the world had *one* team active in this direction?

278   W. ROSS ASHBY

REFERENCES

1. W. ROSS ASHBY, The physical origin of adaptation by trial and error, *J. Gen. Psychol.* **32**, pp. 13–25 (1945).
2. W. ROSS ASHBY, Principles of the self-organizing dynamic system. *J. Gen. Psychol.* **37**, pp. 125–8 (1947).
3. W. ROSS ASHBY, *An Introduction to Cybernetics,* Wiley, New York, 3rd imp. (1958, a).
4. W. ROSS ASHBY, Requisite variety and its implications for the control of complex systems, *Cybernetica,* **1**, pp. 83–99 (1958, b).
5. W. ROSS ASHBY, The mechanism of habituation. In: *The Mechanization of thought Processes.* (Natl. Phys. Lab. Symposium No. 10) H.M.S.O., London (1960).
6. W. ROSS ASHBY, Computers and decision-making, *New Scientist,* 7, p. 746 1960, b).
7. W. ROSS ASHBY, The brain as regulator, *Nature, Lond.* **186**, p. 413 (1960, c).
8. W. ROSS ASHBY, *Design for a Brain; the Origin of Adaptive Behavior,* Wiley, New York, 2nd ed. (1960, d).
9. L. VON BERTALANFFY, An outline of general system theory, *Brit. J. Phil. Sci.* **1**, pp. 134–65 (1950).
10. R. L. BEURLE, Properties of a mass of cells capable of regenerating pulses, *Proc. Roy. Soc.* **B240**, pp. 55–94 (1956).
11. W. R. GARNER and W. J. MCGILL, The relation between information and variance analyses, *Psychometrika* **21**, pp. 219–28 (1956).
12. R. C. JEFFREY, Some recent simplifications of the theory of finite automata. Technical Report 219, Research Laboratory of Electronics, Massachusetts Institute of Technology (27 May 1959).
13. H. S. JENNINGS, *Behavior of the Lower Organisms,* New York (1906).
14. A. J. LOTKA, *Elements of Physical Biology,* Williams & Wilkins, Baltimore (1925).
15. J. G. MARCH and J. A. SIMON, *Organizations,* Wiley, New York (1958).
16. K. H. PRIBRAM, Fifteenth International Congress of Psychology, Brussels (1957).
17. C. E. SHANNON and W. WEAVER, *The Mathematical Theory of Communication,* University of Illinois Press, Urbana (1949).
18. G. SOMMERHOFF, *Analytical Biology,* Oxford University Press, London (1950).

6. General systems theory: The skeleton of science
Kenneth E. Boulding

Reprinted by permission. Copyright 2004 INFORMS. Boulding, K. E. (1956). "General systems theory - the skeleton of science," Management Science, 2: 197-208. The Institute of Management Sciences, now the Institute for Operations Research and the Management Sciences, 901 Elkridge Landing Road, Suite 400, Linthicum, Maryland 21090, USA.

This particular classical paper was written by Kenneth E. Boulding back in 1956, and published in one of the earliest issues of *Management Science* which is currently celebrating its fiftieth anniversary (Hopp, 2004). Congratulations to the team at *Management Science*.

Boulding is a peer of a number of great systems thinkers that introduced and developed the general systems movement in the early fifties. Such thinkers include Ludwig von Bertalanffy, Talcott Parsons, C. West Churchman, Alfred Emerson, Anatol Rapoport, and many more - it is likely that selected writings from these thinkers will appear in future issues of *E:CO*.

For those readers not familiar with the general systems movement (from which complexity thinking arguably emerged) Boulding starts his paper with a brief description:

"General Systems Theory is a name which has come into use to describe a level of theoretical model-building which lies somewhere between the highly generalized constructions of pure mathematics and the specific theories of the specialized disciplines."

This description of GST is very important as many complexity theorists still talk of a theory of complexity, or of a theory of management as if all the complexities and ambiguities of our perceived realities could somehow be reduced to a neat little theoretical package much akin to the physicists' quest for a theory of everything, or as Boulding puts it a "general theory of practically everything". In Boulding's mind GST was to be a tool that would enable mankind to effectively move back and forth between the perfectly describable Platonic world of theory and the fuzzy world of practice. Boulding rightly points out that any claims to any sort of theory of everything are misguided as "[s]uch a theory would be almost without content, for we always pay for generality by sacrificing content, and all we can say about practically everything is almost nothing."

Boulding's *General systems theory* is a sort of manifesto for the systems movement, much of which can be seen to be valid for complex systems theory today. A major role for any GST was to facilitate communication between disparate fields of interest, i.e., to provide a common language with which to discuss systemic problems. A lexicon of complexity science is also emerging, containing concepts such as *emergence, self-organization, chaos, bifurcation, exaptation*, etc. (some of which were also contained in the GST lexicon), which also aims to facilitate cross-disciplinary dialogue (though I personally doubt whether such an all-embracing way to express complexity is possible - there are an infinity of ways to talk about complexity and all of them should be allowed, initially at least).

The modern complexity movement is in some ways quite different from the general systems movement (although to many writers the two seem almost synonymous), but there is a lot to be learnt from the journey general systems theory has taken. Complex systems thinkers share a lot of the aims and ambitions of the original general systems movement, such as the need for cross-disciplinary communication and the development of analytical tools and processes to interact with, and intervene in, a modern complex (systemic) world. In this paper Boulding not only describes the need and role of a general systems framework but also offers a skeleton of what that framework might look like. Some readers may be surprised as to how fresh this paper still is.

Kurt A. Richardson

Hopp, W. J. (2004). "Fifty Years of *Management Science*," *Management Science*, 50(1): 1-7.

VOLUME 2

NUMBER 3

April 1956

Management Science

GENERAL SYSTEMS THEORY—THE SKELETON OF SCIENCE

KENNETH E. BOULDING

University of Michigan

General Systems Theory[1] is a name which has come into use to describe a level of theoretical model-building which lies somewhere between the highly generalized constructions of pure mathematics and the specific theories of the specialized disciplines. Mathematics attempts to organize highly general relationships into a coherent system, a system however which does not have any necessary connections with the "real" world around us. It studies all thinkable relationships abstracted from any concrete situation or body of empirical knowledge. It is not even confined to "quantitative" relationships narrowly defined—indeed, the developments of a mathematics of quality and structure is already on the way, even though it is not as far advanced as the "classical" mathematics of quantity and number. Nevertheless because in a sense mathematics contains all theories it contains none; it is the language of theory, but it does not give us the content. At the other extreme we have the separate disciplines and sciences, with their separate bodies of theory. Each discipline corresponds to a certain segment of the empirical world, and each develops theories which have particular applicability to its own empirical segment. Physics, Chemistry, Biology, Psychology, Sociology, Economics and so on all carve out for themselves certain elements of the experience of man and develop theories and patterns of activity (research) which yield satisfaction in understanding, and which are appropriate to their special segments.

In recent years increasing need has been felt for a body of systematic theoretical constructs which will discuss the general relationships of the empirical world. This is the quest of General Systems Theory. It does not seek, of course, to establish a single, self-contained "general theory of practically everything" which will replace all the special theories of particular disciplines. Such a theory would be almost without content, for we always pay for generality by sacrificing content, and all we can say about practically everything is almost nothing. Somewhere however between the specific that has *no* meaning and the general that has no content there must be, for each purpose and at each level of abstrac-

[1] The name and many of the ideas are to be credited to L. von Bertalanffy, who is not, however, to be held accountable for the ideas of the present author! For a general discussion of Bertalanffy's ideas see *General System Theory: A New Approach to Unity of Science, Human Biology*, Dec., 1951, Vol. 23, p. 303-361.

197

tion, an optimum degree of generality. It is the contention of the General Systems Theorists that this optimum degree of generality in theory is not always reached by the particular sciences. The objectives of General Systems Theory then can be set out with varying degrees of ambition and confidence. At a low level of ambition but with a high degree of confidence it aims to point out similarities in the theoretical constructions of different disciplines, where these exist, and to develop theoretical models having applicability to at least two different fields of study. At a higher level of ambition, but with perhaps a lower degree of confidence it hopes to develop something like a "spectrum" of theories—a system of systems which may perform the function of a "gestalt" in theoretical construction. Such "gestalts" in special fields have been of great value in directing research towards the gaps which they reveal. Thus the periodic table of elements in chemistry directed research for many decades towards the discovery of unknown elements to fill gaps in the table until the table was completely filled. Similarly a "system of systems" might be of value in directing the attention of theorists towards gaps in theoretical models, and might even be of value in pointing towards methods of filling them.

The need for general systems theory is accentuated by the present sociological situation in science. Knowledge is not something which exists and grows in the abstract. It is a function of human organisms and of social organization. Knowledge, that is to say, is always what somebody knows: the most perfect transcript of knowledge in writing is not knowledge if nobody knows it. Knowledge however grows by the receipt of meaningful information—that is, by the intake of messages by a knower which are capable of reorganizing his knowledge. We will quietly duck the question as to what reorganizations constitute "growth" of knowledge by defining "semantic growth" of knowledge as those reorganizations which can profitably be talked about, in writing or speech, by the Right People. Science, that is to say, is what can be talked about profitably by scientists in their role as scientists. The crisis of science today arises because of the increasing difficulty of such profitable talk among scientists as a whole. Specialization has outrun Trade, communication between the disciples becomes increasingly difficult, and the Republic of Learning is breaking up into isolated subcultures with only tenuous lines of communication between them—a situation which threatens intellectual civil war. The reason for this breakup in the body of knowledge is that in the course of specialization the receptors of information themselves become specialized. Hence physicists only talk to physicists, economists to economists—worse still, nuclear physicists only talk to nuclear physicists and econometricians to econometricians. One wonders sometimes if science will not grind to a stop in an assemblage of walled-in hermits, each mumbling to himself words in a private language that only he can understand. In these days the arts may have beaten the sciences to this desert of mutual unintelligibility, but that may be merely because the swift intuitions of art reach the future faster than the plodding leg work of the scientist. The more science breaks into sub-groups, and the less communication is possible among the disciplines, however, the greater chance there is that the total growth of knowledge is being slowed down by the loss of relevant communications. The spread of specialized deafness means that someone who ought to know something that someone else knows isn't able to find it out for lack of generalized ears.

It is one of the main objectives of General Systems Theory to develop these generalized ears, and by developing a framework of general theory to enable one specialist to catch relevant communications from others. Thus the economist who realizes the strong formal similarity between utility theory in economics and field theory in physics[1] is probably in a better position to learn from the physicists than one who does not. Similarly a specialist who works with the growth concept—whether the crystallographer, the virologist, the cytologist, the physiologist, the psychologist, the sociologist or the economist—will be more sensitive to the contributions of other fields if he is aware of the many similarities of the growth process in widely different empirical fields.

There is not much doubt about the demand for general systems theory under one brand name or another. It is a little more embarrassing to inquire into the supply. Does any of it exist, and if so where? What is the chance of getting more of it, and if so, how? The situation might be described as promising and in ferment, though it is not wholly clear what is being promised or brewed. Something which might be called an "interdisciplinary movement" has been abroad for some time. The first signs of this are usually the development of hybrid disciplines. Thus physical chemistry emerged in the third quarter of the nineteenth century, social psychology in the second quarter of the twentieth. In the physical and biological sciences the list of hybrid disciplines is now quite long—biophysics, biochemistry, astrophysics are all well established. In the social sciences social anthropology is fairly well established, economic psychology and economic sociology are just beginning. There are signs, even, that Political Economy, which died in infancy some hundred years ago, may have a re-birth.

In recent years there has been an additional development of great interest in the form of "multisexual" interdisciplines. The hybrid disciplines, as their hyphenated names indicate, come from two respectable and honest academic parents. The newer interdisciplines have a much more varied and occasionally even obscure ancestry, and result from the reorganization of material from many different fields of study. Cybernetics, for instance, comes out of electrical engineering, neurophysiology, physics, biology, with even a dash of economics. Information theory, which originated in communications engineering, has important applications in many fields stretching from biology to the social sciences. Organization theory comes out of economics, sociology, engineering, physiology, and Management Science itself is an equally multidisciplinary product.

On the more empirical and practical side the interdisciplinary movement is reflected in the development of interdepartmental institutes of many kinds. Some of these find their basis of unity in the empirical field which they study, such as institutes of industrial relations, of public administration, of international

[1] See A. G. Pikler, Utility Theories in Field Physics and Mathematical Economics, *British Journal for the Philosophy of Science*, 1955, Vol. 5, pp. 47 and 303.

200   KENNETH BOULDING

affairs, and so on. Others are organized around the application of a common methodology to many different fields and problems, such as the Survey Research Center and the Group Dynamics Center at the University of Michigan. Even more important than these visible developments, perhaps, though harder to perceive and identify, is a growing dissatisfaction in many departments, especially at the level of graduate study, with the existing traditional theoretical backgrounds for the empirical studies which form the major part of the output of Ph.D. theses. To take but a single example from the field with which I am most familiar. It is traditional for studies of labor relations, money and banking, and foreign investment to come out of departments of economics. Many of the needed theoretical models and frameworks in these fields, however, do not come out of "economic theory" as this is usually taught, but from sociology, social psychology, and cultural anthropology. Students in the department of economics however rarely get a chance to become acquainted with these theoretical models, which may be relevant to their studies, and they become impatient with economic theory, much of which may not be relevant.

It is clear that there is a good deal of interdisciplinary excitement abroad. If this excitement is to be productive, however, it must operate within a certain framework of coherence. It is all too easy for the interdisciplinary to degenerate into the undisciplined. If the interdisciplinary movement, therefore, is not to lose that sense of form and structure which is the "discipline" involved in the various separate disciplines, it should develop a structure of its own. This I conceive to be the great task of general systems theory. For the rest of this paper, therefore, I propose to look at some possible ways in which general systems theory might be structured.

Two possible approaches to the organization of general systems theory suggest themselves, which are to be thought of as complementary rather than competitive, or at least as two roads each of which is worth exploring. The first approach is to look over the empirical universe and to pick out certain general *phenomena* which are found in many different disciplines, and to seek to build up general theoretical models relevant to these phenomena. The second approach is to arrange the empirical fields in a hierarchy of complexity of organization of their basic "individual" or unit of behavior, and to try to develop a level of abstraction appropriate to each.

Some examples of the first approach will serve to clarify it, without pretending to be exhaustive. In almost all disciplines, for instance, we find examples of populations—aggregates of individuals conforming to a common definition, to which individuals are added (born) and subtracted (die) and in which the age of the individual is a relevant and identifiable variable. These populations exhibit dynamic movements of their own, which can frequently be described by fairly simple systems of difference equations. The populations of different species also exhibit dynamic interactions among themselves, as in the theory of Volterra. Models of population change and interaction cut across a great many different fields—ecological systems in biology, capital theory in economics which deals with populations of "goods," social ecology, and even certain problems of statistical mechanics. In all these fields population change, both in absolute numbers and in structure, can be discussed in terms of birth and survival functions relating numbers of births and of deaths in specific age groups to various aspects of the system. In all these fields the interaction of population can be discussed in terms of competitive, complementary, or parasitic relationships among populations of different species, whether the species consist of animals, commodities, social classes or molecules.

Another phenomenon of almost universal significance for all disciplines is that of the interaction of an "individual" of some kind with its environment. Every discipline studies some kind of "individual"—electron, atom, molecule, crystal, virus, cell, plant, animal, man, family, tribe, state, church, firm, corporation, university, and so on. Each of these individuals exhibits "behavior," action, or change, and this behavior is considered to be related in some way to the environment of the individual—that is, with other individuals with which it comes into contact or into some relationship. Each individual is thought of as consisting of a structure or complex of individuals of the order immediately below it—atoms are an arrangement of protons and electrons, molecules of atoms, cells of molecules, plants, animals and men of cells, social organizations of men. The "behavior" of each individual is "explained" by the structure and arrangement of the lower individuals of which it is composed, or by certain principles of equilibrium or homeostasis according to which certain "states" of the individual are "preferred." Behavior is described in terms of the restoration of these preferred states when they are disturbed by changes in the environment.

Another phenomenon of universal significance is growth. Growth theory is in a sense a subdivision of the theory of individual "behavior," growth being one important aspect of behavior. Nevertheless there are important differences between equilibrium theory and growth theory, which perhaps warrant giving growth theory a special category. There is hardly a science in which the growth phenomenon does not have some importance, and though there is a great difference in complexity between the growth of crystals, embryos, and societies, many of the principles and concepts which are important at the lower levels are also illuminating at higher levels. Some growth phenomena can be dealt with in terms of relatively simple population models, the solution of which yields growth curves of single variables. At the more complex levels structural problems become dominant and the complex interrelationships between growth and form are the focus of interest. All growth phenomena are sufficiently alike however to suggest that a general theory of growth is by no means an impossibility.[1]

Another aspect of the theory of the individual and also of interrelationships among individuals which might be singled out for special treatment is the theory of information and communication. The information concept as developed by Shannon has had interesting applications outside its original field of electrical engineering. It is not adequate, of course, to deal with problems involving the semantic level of communication. At the biological level however the informa-

[1] See "Towards a General Theory of Growth" by K. E. Boulding, *Canadian Journal of Economics and Political Science*, 19 Aug. 1953, 326–340.

tion concept may serve to develop general notions of structuredness and abstract measures of organization which give us, as it were, a third basic dimension beyond mass and energy. Communication and information processes are found in a wide variety of empirical situations, and are unquestionably essential in the development of organization, both in the biological and the social world.

These various approaches to general systems through various aspects of the empirical world may lead ultimately to something like a general field theory of the dynamics of action and interaction. This, however, is a long way ahead.

A second possible approach to general systems theory is through the arrangement of theoretical systems and constructs in a hierarchy of complexity, roughly corresponding to the complexity of the "individuals" of the various empirical fields. This approach is more systematic than the first, leading towards a "system of systems." It may not replace the first entirely, however, as there may always be important theoretical concepts and constructs lying outside the systematic framework. I suggest below a possible arrangement of "levels" of theoretical discourse.

(i) The first level is that of the static structure. It might be called the level of *frameworks*. This is the geography and anatomy of the universe—the patterns of electrons around a nucleus, the pattern of atoms in a molecular formula, the arrangement of atoms in a crystal, the anatomy of the gene, the cell, the plant, the animal, the mapping of the earth, the solar system, the astronomical universe. The accurate description of these frameworks is the beginning of organized theoretical knowledge in almost any field, for without accuracy in this description of static relationships no accurate functional or dynamic theory is possible. Thus the Copernican revolution was really the discovery of a new static framework for the solar system which permitted a simpler description of its dynamics.

(ii) The next level of systematic analysis is that of the simple dynamic system with predetermined, necessary motions. This might be called the level of *clockworks*. The solar system itself is of course the great clock of the universe from man's point of view, and the deliciously exact predictions of the astronomers are a testimony to the excellence of the clock which they study. Simple machines such as the lever and the pulley, even quite complicated machines like steam engines and dynamos fall mostly under this category. The greater part of the theoretical structure of physics, chemistry, and even of economics falls into this category. Two special cases might be noted. Simple equilibrium systems really fall into the dynamic category, as every equilibrium system must be considered as a limiting case of a dynamic system, and its stability cannot be determined except from the properties of its parent dynamic system. Stochastic dynamic systems leading to equilibria, for all their complexity, also fall into this group of systems; such is the modern view of the atom and even of the molecule, each position or part of the system being given with a certain degree of probability, the whole nevertheless exhibiting a determinate structure. Two types of analytical method are important here, which we may call, with the usage of the economists, comparative statics and true dynamics. In comparative statics we compare two equilibrium positions of the system under different values for the basic parameters. These equilibrium positions are usually expressed as the solution of a set of simultaneous equations. The method of comparative statics is to compare the solutions when the parameters of the equations are changed. Most simple mechanical problems are solved in this way. In true dynamics on the other hand we exhibit the system as a set of difference or differential equations, which are then solved in the form of an explicit function of each variable with time. Such a system may reach a position of stationary equilibrium, or it may not—there are plenty of examples of explosive dynamic systems, a very simple one being the growth of a sum at compound interest! Most physical and chemical reactions and most social systems do in fact exhibit a tendency to equilibrium— otherwise the world would have exploded or imploded long ago.

(iii) The next level is that of the control mechanism or cybernetic system, which might be nicknamed the level of the *thermostat*. This differs from the simple stable equilibrium system mainly in the fact that the transmission and interpretation of information is an essential part of the system. As a result of this the equilibrium position is not merely determined by the equations of the system, but the system will move to the maintenance of any *given* equilibrium, within limits. Thus the thermostat will maintain *any* temperature at which it can be set; the equilibrium temperature of the system is not determined solely by its equations. The trick here of course is that the essential variable of the dynamic system is the *difference* between an "observed" or "recorded" value of the maintained variable and its "ideal" value. If this difference is not zero the system moves so as to diminish it; thus the furnace sends up heat when the temperature as recorded is "too cold" and is turned off when the recorded temperature is "too hot." The homeostasis model, which is of such importance in physiology, is an example of a cybernetic mechanism, and such mechanisms exist through the whole empirical world of the biologist and the social scientist.

(iv) The fourth level is that of the "open system," or self-maintaining structure. This is the level at which life begins to differentiate itself from not-life: it might be called the level of the *cell*. Something like an open system exists, of course, even in physico-chemical equilibrium systems; atomic structures maintain themselves in the midst of a throughput of electrons, molecular structures maintain themselves in the midst of a throughput of atoms. Flames and rivers likewise are essentially open systems of a very simple kind. As we pass up the scale of complexity of organization towards living systems, however, the property of self-maintenance of structure in the midst of a throughput of material becomes of dominant importance. An atom or a molecule can presumably exist without throughput: the existence of even the simplest living organism is inconceivable without ingestion, excretion and metabolic exchange. Closely connected with the property of self-maintenance is the property of self-reproduction. It may be, indeed, that self-reproduction is a more primitive or "lower level" system than the open system, and that the gene and the virus, for instance, may be able to reproduce themselves without being open systems. It is not perhaps an important question at what point in the scale of increasing complexity "life" begins. What is clear, however, is that by the time we have got to systems which both reproduce themselves and maintain themselves in the midst of a throughput of material and energy, we have something to which it would be hard to deny the title of "life."

(v) The fifth level might be called the genetic-societal level; it is typified by the *plant*, and it dominates the empirical world of the botanist. The outstanding characteristics of these systems are first, a division of labor among cells to form a cell-society with differentiated and mutually dependent parts (roots, leaves, seeds, etc.), and second, a sharp differentiation between the genotype and the phenotype, associated with the phenomenon of equifinal or "blueprinted" growth. At this level there are no highly specialized sense organs and information receptors are diffuse and incapable of much throughput of information—it is doubtful whether a tree can distinguish much more than light from dark, long days from short days, cold from hot.

(vi) As we move upward from the plant world towards the animal kingdom we gradually pass over into a new level, the "animal" level, characterized by increased mobility, teleological behavior, and self-awareness. Here we have the development of specialized information-receptors (eyes, ears, etc.) leading to an enormous increase in the intake of information; we have also a great development of nervous systems, leading ultimately to the brain, as an organizer of the information intake into a knowledge structure or "image". Increasingly as we ascend the scale of animal life, behavior is response not to a specific stimulus but to an "image" or knowledge structure or view of the environment as a whole. This image is of course determined ultimately by information received into the organism; the relation between the receipt of information and the building up of an image however is exceedingly complex. It is not a simple piling up or accumulation of information received, although this frequently happens, but a structuring of information into something essentially different from the information itself. After the image structure is well established most information received produces very little change in the image—it goes through the loose structure, as it were, without hitting it, much as a sub-atomic particle might go through an atom without hitting anything. Sometimes however the information is "captured" by the image and added to it, and sometimes the information hits some kind of a "nucleus" of the image and a reorganization takes place, with far reaching and radical changes in behavior in apparent response to what seems like a very small stimulus. The difficulties in the prediction of the behavior of these systems arises largely because of this intervention of the image between the stimulus and the response.

(vii) The next level is the "human" level, that is of the individual human being considered as a system. In addition to all, or nearly all, of the characteristics of animal systems man possesses self consciousness, which is something different from mere awareness. His image, besides being much more complex than that even of the higher animals, has a self-reflexive quality—he not only knows, but knows that he knows. This property is probably bound up with the phenomenon of language and symbolism. It is the capacity for speech—the ability to produce, absorb, and interpret *symbols*, as opposed to mere signs like the warning cry of an animal—which most clearly marks man off from his humbler brethren. Man is distinguished from the animals also by a much more elaborate image of time and relationship; man is probably the only organization that knows that it dies, that contemplates in its behavior a whole life span, and more than a life span. Man exists not only in time and space but in history, and his behavior is profoundly affected by his view of the time process in which he stands.

(viii) Because of the vital importance for the individual man of symbolic images and behavior based on them it is not easy to separate clearly the level of the individual human organism from the next level, that of social organizations. In spite of the occasional stories of feral children raised by animals, man isolated from his fellows is practically unknown. So essential is the symbolic image in human behavior that one suspects that a truly isolated man would not be "human" in the usually accepted sense, though he would be potentially human. Nevertheless it is convenient for some purposes to distinguish the individual human as a system from the social systems which surround him, and in this sense social organizations may be said to constitute another level of organization. The unit of such systems is not perhaps the person—the individual human as such—but the "role"—that part of the person which is concerned with the organization or situation in question, and it is tempting to define social organizations, or almost any social system, as a set of roles tied together with channels of communication. The interrelations of the role and the person however can never be completely neglected—a square person in a round role may become a little rounder, but he also makes the role squarer, and the perception of a role is affected by the personalities of those who have occupied it in the past. At this level we must concern ourselves with the content and meaning of messages, the nature and dimensions of value systems, the transcription of images into a historical record, the subtle symbolizations of art, music, and poetry, and the complex gamut of human emotion. The empirical universe here is human life and society in all its complexity and richness.

(ix) To complete the structure of systems we should add a final turret for transcendental systems, even if we may be accused at this point of having built Babel to the clouds. There are however the ultimates and absolutes and the inescapable unknowables, and they also exhibit systematic structure and relationship. It will be a sad day for man when nobody is allowed to ask questions that do not have any answers.

One advantage of exhibiting a hierarchy of systems in this way is that it gives us some idea of the present gaps in both theoretical and empirical knowledge. Adequate theoretical models extend up to about the fourth level, and not much beyond. Empirical knowledge is deficient at practically all levels. Thus at the level of the static structure, fairly adequate descriptive models are available for geography, chemistry, geology, anatomy, and descriptive social science. Even at this simplest level, however, the problem of the adequate description of complex structures is still far from solved. The theory of indexing and cataloging, for instance, is only in its infancy. Librarians are fairly good at cataloguing books, chemists have begun to catalogue structural formulae, and anthropologists have begun to catalogue culture trails. The cataloguing of events, ideas, theories, statistics, and empirical data has hardly begun. The very multiplication of records however as time goes on will force us into much more adequate cataloguing and reference systems than we now have. This is perhaps the major unsolved theoretical problem at the level of the static structure. In the empirical field there are still great areas where static structures are very imperfectly known, although knowlege is advancing rapidly, thanks to new probing devices such as the electron microscope. The anatomy of that part of the empirical world which lies between the large molecule and the cell however, is still obscure at many points. It is precisely this area however—which includes, for instance, the gene and the virus—that holds the secret of life, and until its anatomy is made clear the nature of the functional systems which are involved will inevitably be obscure.

The level of the "clockwork" is the level of "classical" natural science, especially physics and astronomy, and is probably the most completely developed level in the present state of knowledge, especially if we extend the concept to include the field theory and stochastic models of modern physics. Even here however there are important gaps, especially at the higher empirical levels. There is much yet to be known about the sheer mechanics of cells and nervous systems, of brains and of societies.

Beyond the second level adequate theoretical models get scarcer. The last few years have seen great developments at the third and fourth levels. The theory of control mechanisms ("thermostats") has established itself as the new discipline or cybernetics, and the theory of self-maintaining systems or "open systems" likewise has made rapid strides. We could hardly maintain however that much more than a beginning had been made in these fields. We know very little about the cybernetics of genes and genetic systems, for instance, and still less about the control mechanisms involved in the mental and social world. Similarly the processes of self-maintenance remain essentially mysterious at many points, and although the theoretical possibility of constructing a self-maintaining machine which would be a true open system has been suggested, we seem to be a long way from the actual construction of such a mechanical similitude of life.

Beyond the fourth level it may be doubted whether we have as yet even the rudiments of theoretical systems. The intricate machinery of growth by which the genetic complex organizes the matter around it is almost a complete mystery. Up to now, whatever the future may hold, only God can make a tree. In the face of living systems we are almost helpless; we can occasionally cooperate with systems which we do not understand: we cannot even begin to reproduce them. The ambiguous status of medicine, hovering as it does uneasily between magic and science, is a testimony to the state of systematic knowledge in this area. As we move up the scale the absence of the appropriate theoretical systems becomes ever more noticeable. We can hardly conceive ourselves constructing a system which would be in any recognizable sense "aware," much less self conscious. Nevertheless as we move towards the human and societal level a curious thing happens: the fact that we have, as it were, an inside track, and that we ourselves *are* the systems which we are studying, enables us to utilize systems which we do not really understand. It is almost inconceivable that we should make a machine that would make a poem: nevertheless, poems *are* made by fools like us by processes which are largely hidden from us. The kind of knowledge and skill that we have at the symbolic level is very different from that which we have at lower levels—it is like, shall we say, the "knowhow" of the gene as compared with the knowhow of the biologist. Nevertheless it is a real kind of knowledge and it is the source of the creative achievements of man as artist, writer, architect, and composer.

Perhaps one of the most valuable uses of the above scheme is to prevent us from accepting as final a level of theoretical analysis which is below the level of the empirical world which we are investigating. Because, in a sense, each level incorporates all those below it, much valuable information and insights can be obtained by applying low-level systems to high-level subject matter. Thus most of the theoretical schemes of the social sciences are still at level (ii), just rising now to (iii), although the subject matter clearly involves level (viii). Economics, for instance, is still largely a "mechanics of utility and self interest," in Jevons' masterly phrase. Its theoretical and mathematical base is drawn largely from the level of simple equilibrium theory and dynamic mechanisms. It has hardly begun to use concepts such as information which are appropriate at level (iii), and makes no use of higher level systems. Furthermore, with this crude apparatus it has achieved a modicum of success, in the sense that anybody trying to manipulate an economic system is almost certain to be better off if he knows some economics than if he doesn't. Nevertheless at some point progress in economics is going to depend on its ability to break out of these low-level systems, useful as they are as first approximations, and utilize systems which are more directly appropriate to its universe—when, of course, these systems are discovered. Many other examples could be given—the wholly inappropriate use in psychoanalytic theory, for instance, of the concept of energy, and the long inability of psychology to break loose from a sterile stimulus-response model.

Finally, the above scheme might serve as a mild word of warning even to Management Science. This new discipline represents an important breakaway from overly simple mechanical models in the theory of organization and control. Its emphasis on communication systems and organizational structure, on principles of homeostasis and growth, on decision processes under uncertainty, is carrying us far beyond the simple models of maximizing behavior of even ten years ago. This advance in the level of theoretical analysis is bound to lead to more powerful and fruitful systems. Nevertheless we must never quite forget that even these advances do not carry us much beyond the third and fourth levels, and that in dealing with human personalities and organizations we are dealing with systems in the empirical world far beyond our ability to formulate. We should not be wholly surprised, therefore, if our simpler systems, for all their importance and validity, occasionally let us down.

I chose the subtitle of my paper with some eye to its possible overtones of meaning. General Systems Theory is the skeleton of science in the sense that it aims to provide a framework or structure of systems on which to hang the flesh and blood of particular disciplines and particular subject matters in an orderly and coherent corpus of knowledge. It is also, however, something of a skeleton in a cupboard—the cupboard in this case being the unwillingness of science to admit the very low level of its successes in systematization, and its tendency to shut the door on problems and subject matters which do not fit easily into simple mechanical schemes. Science, for all its successes, still has a very long way to go. General Systems Theory may at times be an embarrassment in pointing out how very far we still have to go, and in deflating excessive philosophical claims for overly simple systems. It also may be helpful however in pointing out to some extent *where* we have to go. The skeleton must come out of the cupboard before its dry bones can live.

7. Science and complexity
Warren Weaver

Originally published as Weaver, W. (1948). "Science and complexity," in *American Scientist*, 36: 536-544. Reproduced with permission. The editors would also like to express their sincere thanks to Mia Smith of American Scientist for providing a high quality digital scan of the original publication.

I t is easy to get caught up in the excitement surrounding the study of complexity and how our new learning might be applied to the problems we face today. We often feel like pioneers in a new land, making new discoveries. For those involved in charting such a course, it is easy to lose historical perspective and the path already taken by others. It is to these earlier pioneers that the Classical Papers Section is dedicated. Such a side trip to the archives can quickly bring the reader a dose of reality, that some "new" ideas are really only "rediscovered." Similarly, our view of the future can gain some perspective when reading about earlier predictions of the future, what we now call the present.

Reaching back almost 60 years, *E:CO* readers are invited to read a classic article by Warren Weaver (1894-1978). For historical setting, this article was published shortly after World War II and is influenced by operations research and the first computers developed for the war effort. During the war, Weaver headed the Applied Mathematics Panel (AAAS, 2004), a position that led to familiarity with many of the top scientists of the era. It was a time of great advances in science and optimism for more growth in the future. This article was also written at the time Weaver was formulating ideas that would later be published with Claude Shannon in *The mathematical theory of communication*, which laid the foundation for information theory. Weaver's thoughts during this time on how computers might be employed in machine translation were later collected in his famous memorandum on the topic that "formulated goals and methods before most people had any idea of what computers might be capable of" (Griffin, 1997).

The optimistic attitude of the power of science is also reflected in "Science and Complexity." In the first part of the article, Weaver offers a historical perspective of problems addressed by science, a classification that separates simple, few-variable problems from the "disorganized complexity" of numerous-variable problems suitable for probability analysis. The problems in the middle are "organized complexity" with a moderate number of variables and interrelationships that cannot be fully captured in probability statistics nor sufficiently reduced to a simple formula.

The second part of the article addresses how the study of organized complexity might be approached. The answer is through harnessing the power of computers and cross-discipline collaboration. Weaver predicts:

"Some scientists will seek and develop for themselves new kinds of collaborative arrangements; that these groups will have members drawn from essentially all fields of science; and that these new ways of working, effectively instrumented by huge computers, will contribute greatly to the advance which the next half century will surely achieve in handling the complex, but essentially organic, problems of the biological and social sciences." (Weaver, 1948)

When reading this, there is a bit of *déjà vu* in what we sometimes hear today of our study of complexity. So too in the statement that "science has, to date, succeeded in solving a bewildering number of relatively easy problems, whereas the hard problems, and the ones which perhaps promise most for man's future, lie ahead" (Weaver, 1948). In the end the reader is left with conflicting feelings of surprise that we are not further along in our understanding of complexity given Weaver's ideas nearly 60 years ago, while also still being optimistic in our success for the same reasons Weaver was optimistic.

Ross Wirth

References

AAAS (2004). AAS resolution: In memoriam: Warren Weaver, 1894-1978. [WWW document]. URL: http://archives.aaas.org/docs/resolutions. php?doc_id=339.

Griffin, E. (1997). "Information theory of Claude Shannon & Warren Weaver," in *A first look at communication theory* (3rd ed.), chapter 4. URL: http://www.afirstlook.com/archive/information.cfm?source=archther

Infoplease (2004). Weaver, Warren. [WWW document]. URL: http://www.infoplease.com/ce6/people/A0851711.html.

UnivIL (2004). Claude E. Shannon and Warren Weaver / The mathematical theory of communication. [WWW document]. URL: http://www.press.uillinois.edu/s99/shannon.html.

SCIENCE AND COMPLEXITY

By WARREN WEAVER

Rockefeller Foundation, New York City

S CIENCE has led to a multitude of results that affect men's lives. Some of these results are embodied in mere conveniences of a relatively trivial sort. Many of them, based on science and developed through technology, are essential to the machinery of modern life. Many other results, especially those associated with the biological and medical sciences, are of unquestioned benefit and comfort. Certain aspects of science have profoundly influenced men's ideas and even their ideals. Still other aspects of science are thoroughly awesome.

How can we get a view of the function that science should have in the developing future of man? How can we appreciate what science really is and, equally important, what science is not? It is, of course, possible to discuss the nature of science in general philosophical terms. For some purposes such a discussion is important and necessary, but for the present a more direct approach is desirable. Let us, as a very realistic politician used to say, let us look at the record. Neglecting the older history of science, we shall go back only three and a half centuries and take a broad view that tries to see the main features, and omits minor details. Let us begin with the physical sciences, rather than the biological, for the place of the life sciences in the descriptive scheme will gradually become evident.

Problems of Simplicity

Speaking roughly, it may be said that the seventeenth, eighteenth, and nineteenth centuries formed the period in which physical science learned variables, which brought us the telephone and the radio, the automobile and the airplane, the phonograph and the moving pictures, the turbine and the Diesel engine, and the modern hydroelectric power plant.

The concurrent progress in biology and medicine was also impressive, but that was of a different character. The significant problems of living organisms are seldom those in which one can rigidly maintain constant all but two variables. Living things are more likely to present situations in which a half-dozen, or even several dozen quantities are all varying simultaneously, and in subtly interconnected ways. Often they present situations in which the essentially important quantities are either non-quantitative, or have at any rate eluded identification or measurement up to the moment. Thus biological and medical problems often involve the consideration of a most complexly organized whole. It is not surprising that up to 1900 the life sciences were largely concerned with the necessary preliminary stages in the application of the scientific method—preliminary stages which chiefly involve collection, description, classification, and the observation of concurrent and apparently correlated

Based upon material presented in Chapter 1, "The Scientists Speak," Boni & Gaer, Inc., 1947. All rights reserved.

Science and Complexity   537

effects. They had only made the brave beginnings of quantitative theories, and hardly even begun detailed explanations of the physical and chemical mechanisms underlying or making up biological events.

To sum up, physical science before 1900 was largely concerned with two-variable *problems of simplicity;* whereas the life sciences, in which these problems of simplicity are not so often significant, had not yet become highly quantitative or analytical in character.

Problems of Disorganized Complexity

Subsequent to 1900 and actually earlier, if one includes heroic pioneers such as Josiah Willard Gibbs, the physical sciences developed an attack on nature of an essentially and dramatically new kind. Rather than study problems which involved two variables or at most three or four, some imaginative minds went to the other extreme, and said: "Let us develop analytical methods which can deal with two billion variables." That is to say, the physical scientists, with the mathematicians often in the vanguard, developed powerful techniques of probability theory and of statistical mechanics to deal with what may be called problems of *disorganized complexity.*

This last phrase calls for explanation. Consider first a simple illustration in order to get the flavor of the idea. The classical dynamics of the nineteenth century was well suited for analyzing and predicting the motion of a single ivory ball as it moves about on a billiard table. In fact, the relationship between positions of the ball and the times at which it reaches these positions forms a typical nineteenth-century problem of simplicity. One can, but with a surprising increase in difficulty, analyze the motion of two or even of three balls on a billiard table. There has been, in fact, considerable study of the mechanics of the standard game of billiards. But, as soon as one tries to analyze the motion of ten or fifteen balls on the table at once, as in pool, the problem becomes unmanageable, not because there is any theoretical difficulty, but just because the actual labor of dealing in specific detail with so many variables turns out to be impracticable.

Imagine, however, a large billiard table with millions of balls rolling over its surface, colliding with one another and with the side rails. The great surprise is that the problem now becomes easier, for the methods of statistical mechanics are applicable. To be sure the detailed history of one special ball can not be traced, but certain important questions can be answered with useful precision, such as: On the average how many balls per second hit a given stretch of rail? On the average how far does a ball move before it is hit by some other ball? On the average how many impacts per second does a ball experience?

Earlier it was stated that the new statistical methods were applicable to problems of disorganized complexity. How does the word "disorganized" apply to the large billiard table with the many balls? It applies because the methods of statistical mechanics are valid only when the balls are distributed, in their positions and motions, in a helter-skelter,

154

that is to say a disorganized, way. For example, the statistical methods would not apply if someone were to arrange the balls in a row parallel to one side rail of the table, and then start them all moving in precisely parallel paths perpendicular to the row in which they stand. Then the balls would never collide with each other nor with two of the rails, and one would not have a situation of disorganized complexity.

From this illustration it is clear what is meant by a problem of disorganized complexity. It is a problem in which the number of variables is very large, and one in which each of the many variables has a behavior which is individually erratic, or perhaps totally unknown. However, in spite of this helter-skelter, or unknown, behavior of all the individual variables, the system as a whole possesses certain orderly and analyzable average properties.

A wide range of experience comes under the label of disorganized complexity. The method applies with increasing precision when the number of variables increases. It applies with entirely useful precision to the experience of a large telephone exchange, in predicting the average frequency of calls, the probability of overlapping calls of the same number, etc. It makes possible the financial stability of a life insurance company. Although the company can have no knowledge whatsoever concerning the approaching death of any one individual, it has dependable knowledge of the average frequency with which deaths will occur.

This last point is interesting and important. Statistical techniques are not restricted to situations where the scientific theory of the individual events is very well known, as in the billiard example where there is a beautifully precise theory for the impact of one ball on another. This technique can also be applied to situations, like the insurance example, where the individual event is as shrouded in mystery as is the chain of complicated and unpredictable events associated with the accidental death of a healthy man.

The examples of the telephone and insurance companies suggests a whole array of practical applications of statistical techniques based on disorganized complexity. In a sense they are unfortunate examples, for they tend to draw attention away from the more fundamental use which science makes of these new techniques. The motions of the atoms which form all matter, as well as the motions of the stars which form the universe, come under the range of these new techniques. The fundamental laws of heredity are analyzed by them. The laws of thermodynamics, which describe basic and inevitable tendencies of all physical systems, are derived from statistical considerations. The entire structure of modern physics, our present concept of the nature of the physical universe, and of the accessible experimental facts concerning it rest on these statistical concepts. Indeed, the whole question of evidence and the way in which knowledge can be inferred from evidence are now recognized to depend on these same statistical ideas, so that probability notions are essential to any theory of knowledge itself.

Science and Complexity 539

Problems of Organized Complexity

This new method of dealing with disorganized complexity, so powerful an advance over the earlier two-variable methods, leaves a great field untouched. One is tempted to oversimplify, and say that scientific methodology went from one extreme to the other—from two variables to an astronomical number—and left untouched a great middle region. The importance of this middle region, moreover, does not depend primarily on the fact that the number of variables involved is moderate—large compared to two, but small compared to the number of atoms in a pinch of salt. The problems in this middle region, in fact, will often involve a considerable number of variables. The really important characteristic of the problems of this middle region, which science has as yet little explored or conquered, lies in the fact that these problems, as contrasted with the disorganized situations with which statistics can cope, show the essential feature of *organization*. In fact, one can refer to this group of problems as those of *organized complexity*.

What makes an evening primrose open when it does? Why does salt water fail to satisfy thirst? Why can one particular genetic strain of microorganism synthesize within its minute body certain organic compounds that another strain of the same organism cannot manufacture? Why is one chemical substance a poison when another, whose molecules have just the same atoms but assembled into a mirror-image pattern, is completely harmless? Why does the amount of manganese in the diet affect the maternal instinct of an animal? What is the description of aging in biochemical terms? What meaning is to be assigned to the question: Is a virus a living organism? What is a gene, and how does the original genetic constitution of a living organism express itself in the developed characteristics of the adult? Do complex protein molecules "know how" to reduplicate their pattern, and is this an essential clue to the problem of reproduction of living creatures? All these are certainly complex problems, but they are not problems of disorganized complexity, to which statistical methods hold the key. They are all problems which involve dealing simultaneously with a *sizable number of factors which are interrelated into an organic whole*. They are all, in the language here proposed, problems of *organized complexity*.

On what does the price of wheat depend? This too is a problem of organized complexity. A very substantial number of relevant variables is involved here, and they are all interrelated in a complicated, but nevertheless not in helter-skelter, fashion.

How can currency be wisely and effectively stabilized? To what extent is it safe to depend on the free interplay of such economic forces as supply and demand? To what extent must systems of economic control be employed to prevent the wide swings from prosperity to depression? These are also obviously complex problems, and they too involve analyzing systems which are organic wholes, with their parts in close interrelation.

How can one explain the behavior pattern of an organized group of persons such as a labor union, or a group of manufacturers, or a racial minority? There are clearly many factors involved here, but it is equally obvious that here also something more is needed than the mathematics of averages. With a given total of national resources that can be brought to bear, what tactics and strategy will most promptly win a war, or better: what sacrifices of present selfish interest will most effectively contribute to a stable, decent, and peaceful world?

These problems—and a wide range of similar problems in the biological, medical, psychological, economic, and political sciences—are just too complicated to yield to the old nineteenth-century techniques which were so dramatically successful on two-, three-, or four-variable problems of simplicity. These new problems, moreover, cannot be handled with the statistical techniques so effective in describing average behavior in problems of disorganized complexity.

These new problems, and the future of the world depends on many of them, requires science to make a third great advance, an advance that must be even greater than the nineteenth-century conquest of problems of simplicity or the twentieth-century victory over problems of disorganized complexity. Science must, over the next 50 years, learn to deal with these problems of organized complexity.

Is there any promise on the horizon that this new advance can really be accomplished? There is much general evidence, and there are two recent instances of especially promising evidence. The general evidence consists in the fact that, in the minds of hundreds of scholars all over the world, important, though necessarily minor, progress is already being made on such problems. As never before, the quantitative experimental methods and the mathematical analytical methods of the physical sciences are being applied to the biological, the medical, and even the social sciences. The results are as yet scattered, but they are highly promising. A good illustration from the life sciences can be seen by a comparison of the present situation in cancer research with what it was twenty-five years ago. It is doubtless true that we are only scratching the surface of the cancer problem, but at least there are now some tools to dig with and there have been located some spots beneath which almost surely there is pay-dirt. We know that certain types of cancer can be induced by certain pure chemicals. Something is known of the inheritance of susceptibility to certain types of cancer. Million-volt rays are available, and the even more intense radiations made possible by atomic physics. There are radioactive isotopes, both for basic studies and for treatment. Scientists are tackling the almost incredibly complicated story of the biochemistry of the aging organism. A base of knowledge concerning the normal cell is being established that makes it possible to recognize and analyze the pathological cell. However distant the goal, we are now at last on the road to a successful solution of this great problem.

In addition to the general growing evidence that problems of organized complexity can be successfully treated, there are at least two promis-

Science and Complexity

ing bits of special evidence. Out of the wickedness of war have come two new developments that may well be of major importance in helping science to solve these complex twentieth-century problems.

The first piece of evidence is the wartime development of new types of electronic computing devices. These devices are, in flexibility and capacity, more like a human brain than like the traditional mechanical computing device of the past. They have "memories" in which vast amounts of information can be stored. They can be "told" to carry out computations of very intricate complexity, and can be left unattended while they go forward automatically with their task. The astounding speed with which they proceed is illustrated by the fact that one small part of such a machine, if set to multiplying two ten-digit numbers, can perform such multiplications some 40,000 times faster than a human operator can say "Jack Robinson." This combination of flexibility, capacity, and speed makes it seem likely that such devices will have a tremendous impact on science. They will make it possible to deal with problems which previously were too complicated, and, more importantly, they will justify and inspire the development of new methods of analysis applicable to these new problems of organized complexity.

The second of the wartime advances is the "mixed-team" approach of operations analysis. These terms require explanation, although they are very familiar to those who were concerned with the application of mathematical methods to military affairs.

As an illustration, consider the over-all problem of convoying troops and supplies across the Atlantic. Take into account the number and effectiveness of the naval vessels available, the character of submarine attacks, and a multitude of other factors, including such an imponderable as the dependability of visual watch when men are tired, sick, or bored. Considering a whole mass of factors, some measurable and some elusive, what procedure would lead to the best over-all plan, that is, best from the combined point of view of speed, safety, cost, and so on? Should the convoys be large or small, fast or slow? Should they zigzag and expose themselves longer to possible attack, or dash in a speedy straight line? How are they to be organized, what defenses are best, and what organization and instruments should be used for watch and attack?

The attempt to answer such broad problems of tactics, or even broader problems of strategy, was the job during the war of certain groups known as the operations analysis groups. Inaugurated with brilliance by the British, the procedure was taken over by this country, and applied with special success in the Navy's anti-submarine campaign and in the Army Air Forces. These operations analysis groups were, moreover, what may be called mixed teams. Although mathematicians, physicists, and engineers were essential, the best of the groups also contained physiologists, biochemists, psychologists, and a variety of representatives of other fields of the biochemical and social sciences. Among the outstanding members of English mixed teams, for example, were an endocrinologist and an X-ray crystallographer. Under the pressure of war, these mixed teams pooled their resources and focused all their different insights on the common problems. It was found, in spite of the modern tendencies toward intense scientific specialization, that members of such diverse groups could work together and could form a unit which was much greater than the mere sum of its parts. It was shown that these groups could tackle certain problems of organized complexity, and get useful answers.

It is tempting to forecast that the great advances that science can and must achieve in the next fifty years will be largely contributed to by voluntary mixed teams, somewhat similar to the operations analysis groups of war days, their activities made effective by the use of large, flexible, and highspeed computing machines. However, it cannot be assumed that this will be the exclusive pattern for future scientific work, for the atmosphere of complete intellectual freedom is essential to science. There will always, and properly, remain those scientists for whom intellectual freedom is necessarily a private affair. Such men must, and should, work alone. Certain deep and imaginative achievements are probably won only in such a way. Variety is, moreover, a proud characteristic of the American way of doing things. Competition between all sorts of methods is good. So there is no intention here to picture a future in which all scientists are organized into set patterns of activity. Not at all. It is merely suggested that some scientists will seek and develop for themselves new kinds of collaborative arrangements; that these groups will have members drawn from essentially all fields of science; and that these new ways of working, effectively instrumented by huge computers, will contribute greatly to the advance which the next half century will surely achieve in handling the complex, but essentially organic, problems of the biological and social sciences.

The Boundaries of Science

Let us return now to our original questions. What is science? What is not science? What may be expected from science?

Science clearly is a way of solving problems—not all problems, but a large class of important and practical ones. The problems with which science can deal are those in which the predominant factors are subject to the basic laws of logic, and are for the most part measurable. Science is a way of organizing reproducible knowledge about such problems; of focusing and disciplining imagination; of weighing evidence; of deciding what is relevant and what is not; of impartially testing hypotheses; of ruthlessly discarding data that prove to be inaccurate or inadequate; of finding, interpreting, and facing facts, and of making the facts of nature the servants of man.

The essence of science is not to be found in its outward appearance, in its physical manifestations; it is to be found in its inner spirit. That austere but exciting technique of inquiry known as the scientific method is what is important about science. This scientific method requires of its practitioners high standards of personal honesty, open-mindedness,

Science and Complexity 543

focused vision, and love of the truth. These are solid virtues, but science has no exclusive lien on them. The poet has these virtues also, and often turns them to higher uses.

Science has made notable progress in its great task of solving logical and quantitative problems. Indeed, the successes have been so numerous and striking, and the failures have been so seldom publicized, that the average man has inevitably come to believe that science is just about the most spectacularly successful enterprise man ever launched. The fact is, of course, that this conclusion is largely justified.

Impressive as the progress has been, science has by no means worked itself out of a job. It is soberly true that science has, to date, succeeded in solving a bewildering number of relatively easy problems, whereas the hard problems, and the ones which perhaps promise most for man's future, lie ahead.

We must, therefore, stop thinking of science in terms of its spectacular successes in solving problems of simplicity. This means, among other things, that we must stop thinking of science in terms of gadgetry. Above all, science must not be thought of as a modern improved black magic capable of accomplishing anything and everything.

Every informed scientist, I think, is confident that science is capable of tremendous further contributions to human welfare. It can continue to go forward in its triumphant march against physical nature, learning new laws, acquiring new power of forecast and control, making new material things for man to use and enjoy. Science can also make further brilliant contributions to our understanding of animate nature, giving men new health and vigor, longer and more effective lives, and a wiser understanding of human behavior. Indeed, I think most informed scientists go even further and expect that the precise, objective, and analytical techniques of science will find useful application in limited areas of the social and political disciplines.

There are even broader claims which can be made for science and the scientific method. As an essential part of his characteristic procedure, the scientist insists on precise definition of terms and clear characterization of his problem. It is easier, of course, to define terms accurately in scientific fields than in many other areas. It remains true, however, that science is an almost overwhelming illustration of the effectiveness of a well-defined and accepted language, a common set of ideas, a common tradition. The way in which this universality has succeeded in cutting across barriers of time and space, across political and cultural boundaries, is highly significant. Perhaps better than in any other intellectual enterprise of man, science has solved the problem of communicating ideas, and has demonstrated the world-wide cooperation and community of interest which then inevitably results.

Yes, science is a powerful tool, and it has an impressive record. But the humble and wise scientist does not expect or hope that science can do everything. He remembers that science teaches respect for special competence, and he does not believe that every social, economic, or

160

political emergency would be automatically dissolved if "the scientists" were only put into control. He does not—with a few aberrant exceptions—expect science to furnish a code of morals, or a basis for esthetics. He does not expect science to furnish the yardstick for measuring, nor the motor for controlling, man's love of beauty and truth, his sense of value, or his convictions of faith. There are rich and essential parts of human life which are alogical, which are immaterial and non-quantitative in character, and which cannot be seen under the microscope, weighed with the balance, nor caught by the most sensitive microphone.

If science deals with quantitative problems of a purely logical character, if science has no recognition of or concern for value or purpose, how can modern scientific man achieve a balanced good life, in which logic is the companion of beauty, and efficiency is the partner of virtue?

In one sense the answer is very simple: our morals must catch up with our machinery. To state the necessity, however, is not to achieve it. The great gap, which lies so forebodingly between our power and our capacity to use power wisely, can only be bridged by a vast combination of efforts. Knowledge of individual and group behavior must be improved. Communication must be improved between peoples of different languages and cultures, as well as between all the varied interests which use the same language, but often with such dangerously differing connotations. A revolutionary advance must be made in our understanding of economic and political factors. Willingness to sacrifice selfish short-term interests, either personal or national, in order to bring about long-term improvement for all must be developed.

None of these advances can be won unless men understand what science really is; all progress must be accomplished in a world in which modern science is an inescapable, ever-expanding influence.

.

Complexity and Organization

8. The architecture of complexity
Herbert Simon

Originally published as Simon, H. (1962). "The architecture of complexity," *Proceedings of the American Philosophical Society*, ISSN 0003-049X, 106(6): 467-482. Reprinted with the kind permission of the American Philosophic Society. Special thanks goes to Mary McDonald.

What is inside and what is on top? Complex systems and hierarchies

In the days - about a decade ago - when a start was made to apply complexity theory to all sorts of real-world problems like social systems and organizations, the notion of 'hierarchy' came under pressure. A number of important insights were responsible for this, including the recognition of the importance of distributed representation, non-local causes, holism and the importance of relationships with two-way communication.

Another more philosophical reason why the notion of hierarchy was resisted had to do with the problem of reductionism. Crude forms of reductionism propose that the world, or systems within the world, are made up of levels arranged in a hierarchical format. Higher level phenomena could then be reduced to physical activity on lower levels. From this perspective the mind, for example, was nothing but the activities of neurons; neurons can be described chemically and chemistry can be reduced to physics. This view is clearly on oversimplification and the resistance to reductionism which followed included a resistance to the notion of hierarchy.

This resistance had specific effects on our thinking about complex systems. They were understood as consisting of components which were all equally important, interacting in a way which undermined the idea of 'central control'. In this phase of complexity studies the influence of chaos theory was still quite strong, and together this resulted, in organizational theory at least, in the notion of 'flat systems'. Organizations, for example, should be seen as things where the resources are distributed throughout the system. A hierarchical understanding of complex systems is just too rigid.

This was certainly an important phase in the development of complexity theory, but more recently it has become clear that this view is restricted in its own way. The main problem is that a view which underplays hierarchy also tends to underplay the fact that complex systems have *structure*. They are not homogenous things. As a matter of fact, it is clear that chaos in itself does not lead to complexity; that structure is an enabling precondition for complexity. The task now is to rethink the notion of structure without simply falling back into a crude form of reductionism.

In facing this task we can return to Herbert Simon's seminal paper from the early 60s. As one reads it, it becomes clear that we could have saved ourselves a lot of trouble by taking Simon seriously. He argues with exceptional clarity for the unavoidability of hierarchies in complex systems. He shows how, from an evolutionary perspective, it is much more efficient for complex systems to be composed of sub-systems which are hierarchically organized. Hierarchy is not an accidental feature of complex systems, it is an essential one.

Of course, complex systems are not *simply* hierarchical systems, and Simon knows this. If they were simply hierarchical, they would be fully decomposable, and, as a result, easy to understand and model. Unfortunately they are not neatly nested like Russian dolls, there are cross-cutting connections. Simon holds the hope that those interactions which do not fit into the overall hierarchy are of less importance and that complex systems are what he calls "nearly decomposable." If he is right, this would mean that our approximations in hierarchical terms would be close enough to the truth to enable a proper understanding.

I think that this assumption is a little optimistic. The cross-cutting connections are nonlinear, and it is therefore difficult to predict their effects in any general way[1]. Perhaps it is better to think of complex systems not as being nearly-decomposable, but as being decomposable and non-decomposable at the same time. These are issues to be worked out in more detail, but what is clear is that we still have to confront the notion of hierarchy in a serious way. In this confrontation Simon's work will be indispensable. Even if the problem of hierarchy does not interest everyone, they should read Simon's paper just for its eloquence and clarity, as well as for the wide range of issues he addresses with insight. I wish more academic papers were written like this!

Paul Cilliers

Notes
[1] I discuss this problem in a little more detail in Cilliers, P. (2001). "Boundaries, hierarchies and networks in complex systems," *International Journal of Innovation Management*, ISSN 1363-9196, 5(2): 135–147.

THE ARCHITECTURE OF COMPLEXITY

HERBERT A. SIMON*

Professor of Administration, Carnegie Institute of Technology

(*Read April 26, 1962*)

A NUMBER of proposals have been advanced in recent years for the development of "general systems theory" which, abstracting from properties peculiar to physical, biological, or social systems, would be applicable to all of them.[1] We might well feel that, while the goal is laudable, systems of such diverse kinds could hardly be expected to have any nontrivial properties in common. Metaphor and analogy can be helpful, or they can be misleading. All depends on whether the similarities the metaphor captures are significant or superficial.

It may not be entirely vain, however, to search for common properties among diverse kinds of complex systems. The ideas that go by the name of cybernetics constitute, if not a theory, at least a point of view that has been proving fruitful over a wide range of applications.[2] It has been useful to look at the behavior of adaptive systems in terms of the concepts of feedback and homeostasis,

and to analyze adaptiveness in terms of the theory of selective information.[3] The ideas of feedback and information provide a frame of reference for viewing a wide range of situations, just as do the ideas of evolution, of relativism, of axiomatic method, and of operationalism.

In this paper I should like to report on some things we have been learning about particular kinds of complex systems encountered in the behavioral sciences. The developments I shall discuss arose in the context of specific phenomena, but the theoretical formulations themselves make little reference to details of structure. Instead they refer primarily to the complexity of the systems under view without specifying the exact content of that complexity. Because of their abstractness, the theories may have relevance—application would be too strong a term—to other kinds of complex systems that are observed in the social, biological, and physical sciences.

In recounting these developments, I shall avoid technical detail, which can generally be found elsewhere. I shall describe each theory in the particular context in which it arose. Then, I shall cite some examples of complex systems, from areas of science other than the initial application, to which the theoretical framework appears relevant. In doing so, I shall make reference to areas of knowledge where I am not expert—perhaps not even literate. I feel quite comfortable in doing so before the members of this society, representing as it does the whole span of the scientific and scholarly endeavor. Collectively you will have little difficulty, I am sure, in distinguishing instances based on idle fancy or sheer ignorance from instances that cast some light on the ways in which complexity exhibits itself wherever it is found in nature. I shall leave to you the final judgment of relevance in your respective fields.

I shall not undertake a formal definition of

* The ideas in this paper have been the topic of many conversations with my colleague, Allen Newell. George W. Corner suggested important improvements in biological content as well as editorial form. I am also indebted, for valuable comments on the manuscript, to Richard H. Meier, John R. Platt, and Warren Weaver. Some of the conjectures about the nearly decomposable structure of the nucleus-atom-molecule hierarchy were checked against the available quantitative data by Andrew Schoene and William Wise. My work in this area has been supported by a Ford Foundation grant for research in organizations and a Carnegie Corporation grant for research on cognitive processes. To all of the above, my warm thanks, and the usual absolution.

[1] See especially the yearbooks of the Society for General Systems Research. Prominent among the exponents of general systems theory are L. von Bertalanffy, K. Boulding, R. W. Gerard, and J. G. Miller. For a more skeptical view—perhaps too skeptical in the light of the present discussion—see H. A. Simon and A. Newell, Models: their uses and limitations, *in* L. D. White, ed., *The state of the social sciences,* 66–83, Chicago, Univ. of Chicago Press, 1956.

[2] N. Wiener, *Cybernetics,* New York, John Wiley & Sons, 1948. For an imaginative forerunner, see A. J. Lotka, *Elements of mathematical biology,* New York, Dover Publications, 1951, first published in 1924 as *Elements of physical biology.*

[3] C. Shannon and W. Weaver, *The mathematical theory of communication,* Urbana, Univ. of Illinois Press, 1949; W. R. Ashby, *Design for a brain,* New York, John Wiley & Sons, 1952.

PROCEEDINGS OF THE AMERICAN PHILOSOPHICAL SOCIETY, VOL. 106, NO. 6, DECEMBER, 1962
467

468   HERBERT A. SIMON   [PROC. AMER. PHIL. SOC.

"complex systems."[4] Roughly, by a complex system I mean one made up of a large number of parts that interact in a nonsimple way. In such systems, the whole is more than the sum of the parts, not in an ultimate, metaphysical sense, but in the important pragmatic sense that, given the properties of the parts and the laws of their interaction, it is not a trivial matter to infer the properties of the whole. In the face of complexity, an in-principle reductionist may be at the same time a pragmatic holist.[5]

The four sections that follow discuss four aspects of complexity. The first offers some comments on the frequency with which complexity takes the form of hierarchy—the complex system being composed of subsystems that, in turn, have their own subsystems, and so on. The second section theorizes about the relation between the structure of a complex system and the time required for it to emerge through evolutionary processes: specifically, it argues that hierarchic systems will evolve far more quickly than non-hierarchic systems of comparable size. The third section explores the dynamic properties of hierarchically-organized systems, and shows how they can be decomposed into subsystems in order to analyze their behavior. The fourth section examines the relation between complex systems and their descriptions.

Thus, the central theme that runs through my remarks is that complexity frequently takes the form of hierarchy, and that hierarchic systems have some common properties that are independent of their specific content. Hierarchy, I shall argue, is one of the central structural schemes that the architect of complexity uses.

---

[4] W. Weaver, in: Science and complexity, *American Scientist* 36: 536, 1948, has distinguished two kinds of complexity, disorganized and organized. We shall be primarily concerned with organized complexity.

[5] See also John R. Platt, Properties of large molecules that go beyond the properties of their chemical sub-groups, *Jour. Theoret. Biol.* 1: 342–358, 1961. Since the reductionism-holism issue is a major *cause de guerre* between scientists and humanists, perhaps we might even hope that peace could be negotiated between the two cultures along the lines of the compromise just suggested. As I go along, I shall have a little to say about complexity in the arts as well as in the natural sciences. I must emphasize the pragmatism of my holism to distinguish it sharply from the position taken by W. M. Elsasser in *The physical foundation of biology*, New York, Pergamon Press, 1958.

HIERARCHIC SYSTEMS

By a *hierarchic system*, or hierarchy, I mean a system that is composed of interrelated subsystems, each of the latter being, in turn, hierarchic in structure until we reach some lowest level of elementary subsystem. In most systems in nature, it is somewhat arbitrary as to where we leave off the partitioning, and what subsystems we take as elementary. Physics makes much use of the concept of "elementary particle" although particles have a disconcerting tendency not to remain elementary very long. Only a couple of generations ago, the atoms themselves were elementary particles; today, to the nuclear physicist they are complex systems. For certain purposes of astronomy, whole stars, or even galaxies, can be regarded as elementary subsystems. In one kind of biological research, a cell may be treated as an elementary subsystem; in another, a protein molecule; in still another, an amino acid residue.

Just why a scientist has a right to treat as elementary a subsystem that is in fact exceedingly complex is one of the questions we shall take up. For the moment, we shall accept the fact that scientists do this all the time, and that if they are careful scientists they usually get away with it.

Etymologically, the word "hierarchy" has had a narrower meaning than I am giving it here. The term has generally been used to refer to a complex system in which each of the subsystems is subordinated by an authority relation to the system it belongs to. More exactly, in a hierarchic formal organization, each system consists of a "boss" and a set of subordinate subsystems. Each of the subsystems has a "boss" who is the immediate subordinate of the boss of the system. We shall want to consider systems in which the relations among subsystems are more complex than in the formal organizational hierarchy just described. We shall want to include systems in which there is no relation of subordination among subsystems. (In fact, even in human organizations, the formal hierarchy exists only on paper; the real flesh-and-blood organization has many inter-part relations other than the lines of formal authority.) For lack of a better term, I shall use hierarchy in the broader sense introduced in the previous paragraphs, to refer to all complex systems analyzable into successive sets of subsystems, and speak of "formal hierarchy" when I want to refer to the more specialized concept.[6]

---

[6] The mathematical term "partitioning" will not do for what I call here a hierarchy; for the set of subsystems,

SOCIAL SYSTEMS

I have already given an example of one kind of hierarchy that is frequently encountered in the social sciences: a formal organization. Business firms, governments, universities all have a clearly visible parts-within-parts structure. But formal organizations are not the only, or even the most common, kind of social hierarchy. Almost all societies have elementary units called families, which may be grouped into villages or tribes, and these into larger groupings, and so on. If we make a chart of social interactions, of who talks to whom, the clusters of dense interaction in the chart will identify a rather well-defined hierarchic structure. The groupings in this structure may be defined operationally by some measure of frequency of interaction in this sociometric matrix.

BIOLOGICAL AND PHYSICAL SYSTEMS

The hierarchical structure of biological systems is a familiar fact. Taking the cell as the building block, we find cells organized into tissues, tissues into organs, organs into systems. Moving downward from the cell, well-defined subsystems—for example, nucleus, cell membrane, microsomes, mitochondria, and so on—have been identified in animal cells.

The hierarchic structure of many physical systems is equally clear-cut. I have already mentioned the two main series. At the microscopic level we have elementary particles, atoms, molecules, macromolecules. At the macroscopic level we have satellite systems, planetary systems, galaxies. Matter is distributed throughout space in a strikingly non-uniform fashion. The most nearly random distributions we find, gases, are not random distributions of elementary particles but random distributions of complex systems, i.e. molecules.

A considerable range of structural types is subsumed under the term hierarchy as I have defined it. By this definition, a diamond is hierarchic, for it is a crystal structure of carbon atoms that can be further decomposed into protons, neutrons, and electrons. However, it is a very "flat" hierarchy, in which the number of first-order subsystems belonging to the crystal can be indefinitely large. A volume of molecular gas is a flat hierarchy in the same sense. In ordinary usage, we

and the successive subsets in each of these defines the partitioning, independently of any systems of relations among the subsets. By hierarchy I mean the partitioning in conjunction with the relations that hold among its parts.

tend to reserve the word hierarchy for a system that is divided into a *small or moderate number* of subsystems, each of which may be further subdivided. Hence, we do not ordinarily think of or refer to a diamond or a gas as a hierarchic structure. Similarly, a linear polymer is simply a chain, which may be very long, of identical subparts, the monomers. At the molecular level it is a very flat hierarchy.

In discussing formal organizations, the number of subordinates who report directly to a single boss is called his *span of control*. I will speak analogously of the *span* of a system, by which I shall mean the number of subsystems into which it is partitioned. Thus, a hierarchic system is flat at a given level if it has a wide span at that level. A diamond has a wide span at the crystal level, but not at the next level down, the molecular level.

In most of our theory construction in the following sections we shall focus our attention on hierarchies of moderate span, but from time to time I shall comment on the extent to which the theories might or might not be expected to apply to very flat hierarchies.

There is one important difference between the physical and biological hierarchies, on the one hand, and social hierarchies, on the other. Most physical and biological hierarchies are described in spatial terms. We detect the organelles in a cell in the way we detect the raisins in a cake—they are "visibly" differentiated substructures localized spatially in the larger structure. On the other hand, we propose to identify social hierarchies not by observing who lives close to whom but by observing who interacts with whom. These two points of view can be reconciled by defining hierarchy in terms of intensity of interaction, but observing that in most biological and physical systems relatively intense interaction implies relative spatial propinquity. One of the interesting characteristics of nerve cells and telephone wires is that they permit very specific strong interactions at great distances. To the extent that interactions are channeled through specialized communications and transportation systems, spatial propinquity becomes less determinative of structure.

SYMBOLIC SYSTEMS

One very important class of systems has been omitted from my examples thus far: systems of human symbolic production. A book is a hierarchy in the sense in which I am using that term. It is generally divided into chapters, the chapters into sections, the sections into paragraphs, the paragraphs into sentences, the sentences into clauses and phrases, the clauses and phrases into words. We may take the words as our elementary units, or further subdivide them, as the linguist often does, into smaller units. If the book is narrative in character, it may divide into "episodes" instead of sections, but divisions there will be.

The hierarchic structure of music, based on such units as movements, parts, themes, phrases, is well known. The hierarchic structure of products of the pictorial arts is more difficult to characterize, but I shall have something to say about it later.

THE EVOLUTION OF COMPLEX SYSTEMS

Let me introduce the topic of evolution with a parable. There once were two watchmakers, named Hora and Tempus, who manufactured very fine watches. Both of them were highly regarded, and the phones in their workshops rang frequently —new customers were constantly calling them. However, Hora prospered, while Tempus became poorer and poorer and finally lost his shop. What was the reason?

The watches the men made consisted of about 1,000 parts each. Tempus had so constructed his that if he had one partly assembled and had to put it down—to answer the phone say—it immediately fell to pieces and had to be reassembled from the elements. The better the customers liked his watches, the more they phoned him, the more difficult it became for him to find enough uninterrupted time to finish a watch.

The watches that Hora made were no less complex than those of Tempus. But he had designed them so that he could put together subassemblies of about ten elements each. Ten of these subassemblies, again, could be put together into a larger subassembly; and a system of ten of the latter subassemblies constituted the whole watch. Hence, when Hora had to put down a partly assembled watch in order to answer the phone, he lost only a small part of his work, and he assembled his watches in only a fraction of the man-hours it took Tempus.

It is rather easy to make a quantitative analysis of the relative difficulty of the tasks of Tempus and Hora: Suppose the probability that an interruption will occur while a part is being added to an incomplete assembly is p. Then the probability that Tempus can complete a watch he has started without interruption is $(1-p)^{1000}$—a very small number unless p is .001 or less. Each interruption will cost, on the average, the time to as-

semble $1/p$ parts (the expected number assembled before interruption). On the other hand, Hora has to complete one hundred eleven sub-assemblies of ten parts each. The probability that he will not be interrupted while completing any one of these is $(1-p)^{10}$, and each interruption will cost only about the time required to assemble five parts.[7]

Now if p is about .01—that is, there is one chance in a hundred that either watchmaker will be interrupted while adding any one part to an assembly—then a straightforward calculation shows that it will take Tempus, on the average, about four thousand times as long to assemble a watch as Hora.

We arrive at the estimate as follows:

1. Hora must make 111 times as many complete assemblies per watch as Tempus; but,

2. Tempus will lose on the average 20 times as much work for each interrupted assembly as Hora [100 parts, on the average, as against 5]; and,

3. Tempus will complete an assembly only 44 times per million attempts ($.99^{1000} = 44 \times 10^{-6}$), while Hora will complete nine out of ten ($.99^{10} = 9 \times 10^{-1}$). Hence Tempus will have to make 20,000 as many attempts per completed assembly as Hora. $(9 \times 10^{-1})/(44 \times 10^{-6}) = 2 \times 10^4$. Multiplying these three ratios, we get:

$$1/111 \times 100/5 \times .99^{10}/.99^{1000}$$
$$= 1/111 \times 20 \times 20,000 \sim 4,000.$$

---

[7] The speculations on speed of evolution were first suggested by H. Jacobson's application of information theory to estimating the time required for biological evolution. See his paper, Information, reproduction, and the origin of life, in *American Scientist* 43: 119–127, January, 1955. From thermodynamic considerations it is possible to estimate the amount of increase in entropy that occurs when a complex system decomposes into its elements. (See, for example, R. B. Setlow and E. C. Pollard, *Molecular biophysics*, 63–65, Reading, Mass., Addison-Wesley Publishing Co., 1962, and references cited there.) But entropy is the logarithm of a probability, hence information, the negative of entropy, can be interpreted as the logarithm of the reciprocal of the probability—the "improbability," so to speak. The essential idea in Jacobson's model is that the expected time required for the system to reach a particular state is inversely proportional to the probability of the state—hence increases exponentially with the amount of information (negentropy) of the state. Following this line of argument, but not introducing the notion of levels and stable subassemblies, Jacobson arrived at estimates of the time required for evolution so large as to make the event rather improbable. Our analysis, carried through in the same way, but with attention to the stable intermediate forms, produces very much smaller estimates.

BIOLOGICAL EVOLUTION

What lessons can we draw from our parable for biological evolution? Let us interpret a partially completed subassembly of k elementary parts as the coexistence of k parts in a small volume—ignoring their relative orientations. The model assumes that parts are entering the volume at a constant rate, but that there is a constant probability, p, that the part will be dispersed before another is added, unless the assembly reaches a stable state. These assumptions are not particularly realistic. They undoubtedly underestimate the decrease in probability of achieving the assembly with increase in the size of the assembly. Hence the assumptions understate—probably by a large factor—the relative advantage of a hierarchic structure.

Although we cannot, therefore, take the numerical estimate seriously the lesson for biological evolution is quite clear and direct. The time required for the evolution of a complex form from simple elements depends critically on the numbers and distribution of potential intermediate stable forms. In particular, if there exists a hierarchy of potential stable "subassemblies," with about the same span, s, at each level of the hierarchy, then the time required for a subassembly can be expected to be about the same at each level—that is proportional to $1/(1-p)^s$. The time required for the assembly of a system of n elements will be proportional to $\log_s n$, that is, to the number of levels in the system. One would say—with more illustrative than literal intent—that the time required for the evolution of multi-celled organisms from single-celled organisms might be of the same order of magnitude as the time required for the evolution of single-celled organisms from macromolecules. The same argument could be applied to the evolution of proteins from amino acids, of molecules from atoms, of atoms from elementary particles.

A whole host of objections to this oversimplified scheme will occur, I am sure, to every working biologist, chemist, and physicist. Before turning to matters I know more about, I shall mention three of these problems, leaving the rest to the attention of the specialists.

First, in spite of the overtones of the watchmaker parable, the theory assumes no teleological mechanism. The complex forms can arise from the simple ones by purely random processes. (I shall propose another model in a moment that shows this clearly.) Direction is provided to the scheme by the stability of the complex forms, once these come into existence. But this is nothing more than survival of the fittest—i.e., of the stable.

Second, not all large systems appear hierarchical. For example, most polymers—e.g., nylon—are simply linear chains of large numbers of identical components, the monomers. However, for present purposes we can simply regard such a structure as a hierarchy with a span of one—the limiting case. For a chain of any length represents a state of relative equilibrium.[8]

Third, the evolution of complex systems from simple elements implies nothing, one way or the other, about the change in entropy of the entire system. If the process absorbs free energy, the complex system will have a smaller entropy than the elements; if it releases free energy, the opposite will be true. The former alternative is the one that holds for most biological systems, and the net inflow of free energy has to be supplied from the sun or some other source if the second law of thermodynamics is not to be violated. For the evolutionary process we are describing, the equilibria of the intermediate states need have only local and not global stability, and they may be stable only in the steady state—that is, as long as there is an external source of free energy that may be drawn upon.[9]

Because organisms are not energetically closed systems, there is no way to deduce the direction, much less the rate, of evolution from classical thermodynamic considerations. All estimates indicate that the amount of entropy, measured in physical units, involved in the formation of a one-celled biological organism is trivially small—about -10^{-11} cal/degree.[10] The "improbability" of evolution has nothing to do with this quantity of entropy, which is produced by every bacterial cell every generation. The irrelevance of quantity of

[8] There is a well-developed theory of polymer size, based on models of random assembly. See for example P. J. Flory, *Principles of polymer chemistry*, ch. 8, Ithaca, Cornell Univ. Press, 1953. Since *all* subassemblies in the polymerization theory are stable, limitation of molecular growth depends on "poisoning" of terminal groups by impurities or formation of cycles rather than upon disruption of partially-formed chains.

[9] This point has been made many times before, but it cannot be emphasized too strongly. For further discussion, see Setlow and Pollard, *op. cit.*, 49–64; E. Schrodinger, *What is life?* Cambridge Univ. Press, 1945; and H. Linschitz, The information content of a bacterial cell, in H. Questler, ed., *Information theory in biology*, 251–262, Urbana, Univ. of Illinois Press, 1953.

[10] See Linschitz, *op. cit.* This quantity, 10^{-11} cal/degree, corresponds to about 10^{13} bits of information.

information, in this sense, to speed of evolution can also be seen from the fact that exactly as much information is required to "copy" a cell through the reproductive process as to produce the first cell through evolution.

The effect of the existence of stable intermediate forms exercises a powerful effect on the evolution of complex forms that may be likened to the dramatic effect of catalysts upon reaction rates and steady state distribution of reaction products in open systems.[11] In neither case does the entropy change provide us with a guide to system behavior.

PROBLEM SOLVING AS NATURAL SELECTION

Let us turn now to some phenomena that have no obvious connection with biological evolution: human problem-solving processes. Consider, for example, the task of discovering the proof for a difficult theorem. The process can be—and often has been—described as a search through a maze. Starting with the axioms and previously proved theorems, various transformations allowed by the rules of the mathematical systems are attempted, to obtain new expressions. These are modified in turn until, with persistence and good fortune, a sequence or path of transformations is discovered that leads to the goal.

The process usually involves a great deal of trial and error. Various paths are tried; some are abandoned, others are pushed further. Before a solution is found, a great many paths of the maze may be explored. The more difficult and novel the problem, the greater is likely to be the amount of trial and error required to find a solution. At the same time, the trial and error is not completely random or blind; it is, in fact, rather highly selective. The new expressions that are obtained by transforming given ones are examined to see whether they represent progress toward the goal. Indications of progress spur further search in the same direction; lack of progress signals the abandonment of a line of search. Problem solving requires *selective* trial and error.[12]

A little reflection reveals that cues signaling progress play the same role in the problem-solving process that stable intermediate forms play in the biological evolutionary process. In fact, we can take over the watchmaker parable and apply it also to problem solving. In problem solving, a partial result that represents recognizable progress toward the goal plays the role of a stable subassembly.

Suppose that the task is to open a safe whose lock has ten dials, each with one hundred possible settings, numbered from 0 to 99. How long will it take to open the safe by a blind trial-and-error search for the correct setting? Since there are 100^{10} possible settings, we may expect to examine about one-half of these, on the average, before finding the correct one—that is, fifty billion billion settings. Suppose, however, that the safe is defective, so that a click can be heard when any one dial is turned to the correct setting. Now each dial can be adjusted independently, and does not need to be touched again while the others are being set. The total number of settings that has to be tried is only 10×50, or five hundred. The task of opening the safe has been altered, by the cues the clicks provide, from a practically impossible one to a trivial one.[13]

A considerable amount has been learned in the past five years about the nature of the mazes that represent common human problem-solving tasks—proving theorems, solving puzzles, playing chess, making investments, balancing assembly lines, to mention a few. All that we have learned about these mazes points to the same conclusion: that human problem solving, from the most blundering to the most insightful, involves nothing more than varying mixtures of trial and error and selectivity. The selectivity derives from various rules of

amplifier, 215–233 in C. E. Shannon and J. McCarthy, *Automata studies*, Princeton, Princeton Univ. Press, 1956.

[13] The clicking safe example was supplied by D. P. Simon. Ashby, *op. cit.*, 230, has called the selectivity involved in situations of this kind "selection by components." The even greater reduction in time produced by hierarchization in the clicking safe example, as compared with the watchmaker's metaphor, is due to the fact that a random *search* for the correct combination is involved in the former case, while in the latter the parts come together in the right order. It is not clear which of these metaphors provides the better model for biological evolution, but we may be sure that the watchmaker's metaphor gives an exceedingly conservative estimate of the savings due to hierarchization. The safe may give an excessively high estimate because it assumes all possible arrangements of the elements to be equally probable.

[11] See H. Kacser, Some physico-chemical aspects of biological organization, Appendix, pp. 191–249 in C. H. Waddington, *The strategy of the genes*, London, George Allen & Unwin, 1957.

[12] See A. Newell, J. C. Shaw, and H. A. Simon, Empirical explorations of the logic theory machine, *Proceedings of the 1957 Western Joint Computer Conference*, February, 1957, New York: Institute of Radio Engineers; Chess-playing programs and the problem of complexity, *IBM Journal of Research and Development* **2**: 320–335, October, 1958; and for a similar view of problem solving, W. R. Ashby, Design for an intelligence

thumb, or heuristics, that suggest which paths should be tried first and which leads are promising. We do not need to postulate processes more sophisticated than those involved in organic evolution to explain how enormous problem mazes are cut down to quite reasonable size.[14]

THE SOURCES OF SELECTIVITY

When we examine the sources from which the problem-solving system, or the evolving system, as the case may be, derives its selectivity, we discover that selectivity can always be equated with some kind of feedback of information from the environment.

Let us consider the case of problem solving first. There are two basic kinds of selectivity. One we have already noted: various paths are tried out, the consequences of following them are noted, and this information is used to guide further search. In the same way, in organic evolution, various complexes come into being, at least evanescently, and those that are stable provide new building blocks for further construction. It is this information about stable configurations, and not free energy or negentropy from the sun, that guides the process of evolution and provides the selectivity that is essential to account for its rapidity.

The second source of selectivity in problem solving is previous experience. We see this particularly clearly when the problem to be solved is similar to one that has been solved before. Then, by simply trying again the paths that led to the earlier solution, or their analogues, trial-and-error search is greatly reduced or altogether eliminated.

What corresponds to this latter kind of information in organic evolution? The closest analogue is reproduction. Once we reach the level of self-reproducing systems, a complex system, when it has once been achieved, can be multiplied indefinitely. Reproduction in fact allows the inheritance of acquired characteristics, but at the level of genetic material, of course; i.e., only characteristics acquired by the genes can be inherited. We shall return to the topic of reproduction in the final section of this paper.

ON EMPIRES AND EMPIRE-BUILDING

We have not exhausted the categories of complex systems to which the watchmaker argument can reasonably be applied. Philip assembled his

[14] A. Newell and H. A. Simon, Computer simulation of human thinking, *Science* 134: 2011-2017, December 22, 1961.

Macedonian empire and gave it to his son, to be later combined with the Persian subassembly and others into Alexander's greater system. On Alexander's death, his empire did not crumble to dust, but fragmented into some of the major subsystems that had composed it.

The watchmaker argument implies that if one would be Alexander, one should be born into a world where large stable political systems already exist. Where this condition was not fulfilled, as on the Scythian and Indian frontiers, Alexander found empire building a slippery business. So too, T. E. Lawrence's organizing of the Arabian revolt against the Turks was limited by the character of his largest stable building blocks, the separate, suspicious desert tribes.

The profession of history places a greater value upon the validated particular fact than upon tendentious generalization. I shall not elaborate upon my fancy, therefore, but will leave it to historians to decide whether anything can be learned for the interpretation of history from an abstract theory of hierarchic complex systems.

CONCLUSION: THE EVOLUTIONARY EXPLANATION OF HIERARCHY

We have shown thus far that complex systems will evolve from simple systems much more rapidly if there are stable intermediate forms than if there are not. The resulting complex forms in the former case will be hierarchic. We have only to turn the argument around to explain the observed predominance of hierarchies among the complex systems nature presents to us. Among possible complex forms, hierarchies are the ones that have the time to evolve. The hypothesis that complexity will be hierarchic makes no distinction among very flat hierarchies, like crystals, and tissues, and polymers, and the intermediate forms. Indeed, in the complex systems we encounter in nature, examples of both forms are prominent. A more complete theory than the one we have developed here would presumably have something to say about the determinants of width of span in these systems.

NEARLY DECOMPOSABLE SYSTEMS

In hierarchic systems, we can distinguish between the interactions *among* subsystems, on the one hand, and the interactions *within* subsystems —i.e., among the parts of those subsystems—on the other. The interactions at the different levels may be, and often will be, of different orders of

	A1	A2	A3	B1	B2	C1	C2	C3
A1	—	100	—	2	—	—	—	—
A2	100	—	100	1	1	—	—	—
A3	—	100	—	—	2	—	—	—
B1	2	1	—	—	100	2	1	—
B2	—	1	2	100	—	—	1	2
C1	—	—	—	2	—	—	100	—
C2	—	—	—	1	—	100	—	100
C3	—	—	—	—	2	—	100	—

FIG. 1. A hypothetical nearly-decomposable system. In terms of the heat-exchange example of the text, A1, A2, and A3 may be interpreted as cubicles in one room, B1 and B2 as cubicles in a second room, and C1, C2, and C3 as cubicles in a third. The matrix entries then are the heat diffusion coefficients between cubicles.

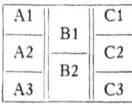

A1		C1
	B1	
A2		C2
	B2	
A3		C3

magnitude. In a formal organization there will generally be more interaction, on the average, between two employees who are members of the same department than between two employees from different departments. In organic substances, intermolecular forces will generally be weaker than molecular forces, and molecular forces than nuclear forces.

In a rare gas, the intermolecular forces will be negligible compared to those binding the molecules —we can treat the individual particles, for many purposes, as if they were independent of each other. We can describe such a system as *decomposable* into the subsystems comprised of the individual particles. As the gas becomes denser, molecular interactions become more significant. But over some range, we can treat the decomposable case as a limit, and as a first approximation. We can use a theory of perfect gases, for example, to describe approximately the behavior of actual gases if they are not too dense. As a second approximation, we may move to a theory of *nearly decomposable* systems, in which the interactions among the subsystems are weak, but not negligible.

At least some kinds of hierarchic systems can be approximated successfully as nearly decomposable systems. The main theoretical findings from the approach can be summed up in two propositions:

(*a*) in a nearly decomposable system, the short-run behavior of each of the component subsystems is approximately independent of the short-run behavior of the other components; (*b*) in the long run, the behavior of any one of the components depends in only an aggregate way on the behavior of the other components.

Let me provide a very concrete simple example of a nearly decomposable system.[15] Consider a building whose outside walls provide perfect thermal insulation from the environment. We shall take these walls as the boundary of our system. The building is divided into a large number of rooms, the walls between them being good, but not perfect, insulators. The walls between rooms are the boundaries of our major subsystems. Each room is divided by partitions into a number of cubicles, but the partitions are poor insulators. A thermometer hangs in each cubicle. Suppose that at the time of our first observation of the system there is a wide variation in temperature from cubicle to cubicle and from room to room—the various cubicles within the building are in a state of thermal disequilibrium. When we take new temperature readings several hours later, what shall we find? There will be very little variation in temperature among the cubicles within each single room, but there may still be large temperature variations *among* rooms. When we take readings again several days later, we find an almost uniform temperature throughout the building; the temperature differences among rooms have virtually disappeared.

We can describe the process of equilibration formally by setting up the usual equations of heat flow. The equations can be represented by the matrix of their coefficients, r_{ij}, where r_{ij} is the rate at which heat flows from the ith cubicle to the jth cubicle per degree difference in their temperatures. If cubicles i and j do not have a common wall, r_{ij} will be zero. If cubicles i and j have a common wall, and are in the same room, r_{ij} will be large. If cubicles i and j are separated by the wall of a

---

15 This discussion of near-decomposability is based upon H. A. Simon and A. Ando, Aggregation of variables in dynamic systems, *Econometrica* **29**: 111–138, April, 1961. The example is drawn from the same source, 117–118. The theory has been further developed and applied to a variety of economic and political phenomena by Ando and F. M. Fisher. See F. M. Fisher, On the cost of approximate specification in simultaneous equation estimation, *Econometrica* **29**: 139–170, April, 1961, and F. M. Fisher and A. Ando, Two theorems on *Ceteris Paribus* in the analysis of dynamic systems, *American Political Science Review* **61**: 103–113, March, 1962.

differentiation of cells in development. It appears more natural to conceptualize that mechanism as based on a process description, and a somewhat more complex interpretive process that produces the adult organism in a sequence of stages, each new stage in development representing the effect of an operator upon the previous one.

It is harder to conceptualize the interrelation of these two descriptions. Interrelated they must be, for enough has been learned of gene-enzyme mechanisms to show that these play a major role in development as in cell metabolism. The single clue we obtain from our earlier discussion is that the description may itself be hierarchical, or nearly decomposable, in structure, the lower levels governing the fast, "high-frequency" dynamics of the individual cell, the higher level interactions governing the slow, "low-frequency" dynamics of the developing multi-cellular organism.

There are only bits of evidence, apart from the facts of recapitulation, that the genetic program is organized in this way, but such evidence as exists is compatible with this notion.[25] To the extent that we can differentiate the genetic information that governs cell metabolism from the genetic information that governs the development of differentiated cells in the multi-cellular organization, we simplify enormously—as we have already seen —our task of theoretical description. But I have perhaps pressed this speculation far enough.

The generalization that in evolving systems whose descriptions are stored in a process language, we might expect ontogeny partially to recapitulate phylogeny has applications outside the

[25] There is considerable evidence that successive genes along a chromosome often determine enzymes controlling successive stages of protein syntheses. For a review of some of this evidence, see P. E. Hartman, Transduction: a comparative review, *in* W. D. McElroy and B. Glass (eds.), *The chemical basis of heredity,* Baltimore, Johns Hopkins Press, 1957, at pp. 442–454. Evidence for differential activity of genes in different tissues and at different stages of development is discussed by J. G. Gall, Chromosomal Differentiation, *in* W. D. McElroy and B. Glass (eds.), *The chemical basis of development,* Baltimore, Johns Hopkins Press, 1958, at pp. 103–135. Finally, a model very like that proposed here has been independently, and far more fully, outlined by J. R. Platt, A 'book model' of genetic information transfer in cells and tissues, *in* Kasha and Pullman (eds.), *Horizons in biochemistry,* New York, Academic Press, forthcoming. Of course, this kind of mechanism is not the only one in which development could be controlled by a process description. Induction, in the form envisaged in Spemann's organizer theory, is based on process description, in which metabolites in already formed tissue control the next stages of development.

realm of biology. It can be applied as readily, for example, to the transmission of knowledge in the educational process. In most subjects, particularly in the rapidly advancing sciences, the progress from elementary to advanced courses is to a considerable extent a progress through the conceptual history of the science itself. Fortunately, the recapitulation is seldom literal—any more than it is in the biological case. We do not teach the phlogiston theory in chemistry in order later to correct it. (I am not sure I could not cite examples in other subjects where we do exactly that.) But curriculum revisions that rid us of the accumulations of the past are infrequent and painful. Nor are they always desirable—partial recapitulation may, in many instances, provide the most expeditious route to advanced knowledge.

SUMMARY: THE DESCRIPTION OF COMPLEXITY

How complex or simple a structure is depends critically upon the way in which we describe it. Most of the complex structures found in the world are enormously redundant, and we can use this redundancy to simplify their description. But to use it, to achieve the simplification, we must find the right representation.

The notion of substituting a process description for a state description of nature has played a central role in the development of modern science. Dynamic laws, expressed in the form of systems of differential or difference equations, have in a large number of cases provided the clue for the simple description of the complex. In the preceding paragraphs I have tried to show that this characteristic of scientific inquiry is not accidental or superficial. The correlation between state description and process description is basic to the functioning of any adaptive organism, to its capacity for acting purposefully upon its environment. Our present-day understanding of genetic mechanisms suggests that even in describing itself the multi-cellular organism finds a process description—a genetically encoded program—to be the parsimonious and useful representation.

CONCLUSION

Our speculations have carried us over a rather alarming array of topics, but that is the price we must pay if we wish to seek properties common to many sorts of complex systems. My thesis has been that one path to the construction of a nontrivial theory of complex systems is by way of a theory of hierarchy. Empirically, a large proportion of the complex systems we observe in nature exhibit hierarchic structure. On theoretical grounds we could expect complex systems to be hierarchies in a world in which complexity had to evolve from simplicity. In their dynamics, hierarchies have a property, near-decomposability, that greatly simplifies their behavior. Near-decomposability also simplifies the description of a complex system, and makes it easier to understand how the information needed for the development or reproduction of the system can be stored in reasonable compass.

In both science and engineering, the study of "systems" is an increasingly popular activity. Its popularity is more a response to a pressing need for synthesizing and analyzing complexity than it is to any large development of a body of knowledge and technique for dealing with complexity. If this popularity is to be more than a fad, necessity will have to mother invention and provide substance to go with the name. The explorations reviewed here represent one particular direction of search for such substance.

9. Futurology and the future of systems analysis

Ida R. Hoos

Originally published in Hoos, I. R. (1972). *Systems analysis in public policy*, Los Angeles, CA: University of California, ISBN 0520021045, Ch. 8, pp. 235-247. Reproduced by kind permission of the University of California.

It is the curse of those who try to create new ideas to be forever stumbling over the same ideas in long forgotten papers. As a defence mechanism we periodically change the language and the boundaries of our disciplines so that we can confidently discount 'out of date' ideas as belonging to a previous age. Ida Hoos's paper was written in the 'Space Age' when systems analysis claimed credit for engineering the Moon landings and RAND's mathematical models of War were still used to shape Strategy. In the 'Information Age' we have matured beyond such naivety and might be tempted to ignore this paper as no longer relevant. We would be wrong to do so. Hoos grapples with the problems of prediction and designing a better future for social systems and, in doing so, presages many of the key ideas of complex systems thinking, albeit in 'Space Age' terms.

To begin Hoos seeks to debunk some of the "normative, methodological, and unsupported assumptions" of the systems approach to futurology. The idea that adequate formal models can be built and sufficiently complete and future-proof data banks accumulated so as to allow "comprehensive anticipatory design science" is challenged by the assertion that "the design of the future is little more than an image projection, more revealing of the creator's *Weltanschauung* than of the form and direction of social changes ahead." The paper also explores the "basic philosophical conflict between free will and determinism" that lies at the heart of attempts to conceive and operate models of the future as a basis for design and social policy. Social engineering can readily become self-fulfilling prophecy, all the more potent and dangerous for being cast in the mould of 'scientific' and 'rational' management.

Systems analysis, when applied prospectively, depends upon an ability to study the future with tools that the analysts themselves recognize to be "crude" and is sold on the basis that "the future designed through rational procedures will be better than its much maligned, disorderly, democratic alternative, arrived at through the presumed anarchy of social forces without vector." In this context science supports the egocentric fantasy that Man is capable of faithfully apprehending and understanding the world and (even more fantastical) controlling it. The systems approach, while purporting to be holistic, actually delivers a deeply reductionist result. There are distinct anticipations of the ideas of complexity thinking in Hoos's suggestion that "experience has shown that the orderly and predictable factors may, in the final analysis, be those of least importance in the dynamics and direction of social change."

But the paper is much more than a philosophical treatise. It engages a clear sight upon the social and political realities of applying systems thinking to the future. The implications of public policy on research grants and the social kudos given to approaches that are (or appear) systematic and technically sophisticated are called out as having diverted executive attention to the "wrong questions."

At times Hoos's writing is florid (e.g., "Deeper probe reveals how thin lies the veneer of glossolalia over fuzzy conceptualization and hyperkinetic data accumulation.") and in the second half of the paper the style becomes rather too polemical, even vitriolic in tone, with a correlated disappearance of external references, indicating an underlying 'agenda' beyond the merely intellectual. There is also much with which to disagree - especially for one whose career is built on internal consultancy within a government department - but a rich scattering of intellectual gems makes the paper's sins forgivable and rewards the reader who is seriously seeking insight rather than a shallow confirmation of prior beliefs.

This Space Age paper raises important questions about the role and efficacy of systems thinking as an approach to support executive decision-making and management, concluding that the techniques of systems analysis are not "appropriate to deal with problems which are essentially human and social" and that "the direction in which they are developing promises little improvement." A generation later, in the Information Age, the questions remain equally valid and represent a real and significant challenge for complex systems thinking to address.

Graham Mathieson

[8]

Futurology and the Future of Systems Analysis

Forecasting the Future through Systems Techniques

Applied in the future tense, systems analysis takes on new and portentous proportions. When used as the methodology by which to "study," "design," or "forecast" the future, its techniques carry with them all the pitfalls and shortcomings of their applications in the present and the added difficulties attendant on "studying" something that has not yet happened. Lacking knowledge of what is yet to come, social forecasters attempt to achieve a future perfect state by devising a "rational" plan even though, as we have seen, their methods have not contributed demonstrably to a better present. In their endeavor they are, nonetheless, encouraged and abetted by all those who yearn for the orderly future as relief from the chaotic present. For the systems entrepreneur in search of new markets for his reservoir of restless talents, futurism invites unfettered play of imagination.

Just as systems analysis of the conventional type offered an enticing grab bag to practitioners with wide diversity and background, the art of designing the future has attracted an even greater heterogeneity. Arrogating to themselves the task of creating a better world, philosophers, urban planners, sociologists, economists, engineers, and many others have banded together in societies of which the leaders, prone to quote one another reverentially, bask in the glow of mutual adulation. With their conferences proceeding in cybernetic fashion, each one's result causing another to occur, futurists engage in solemn methodological discourse. Uninhibited by time or space, they indulge in simulations

235

that range from the presumptuous to the ludicrous, a description the more apt when one recalls its derivation from Latin, *ludi,* meaning *public games and spectacles.* They posit a supranational model in which nations will behave more rationally than the people who populate them. They blithely overlook the eternal struggle for existence that makes coexistence a chimera. The design of the future "one world" demonstrates even more glaringly the gap between the perfection of the system dreamed up by the international jet set intellectuals and the imperfections which are the down-to-earth realities.

The more ambitious the model, the more likely is the fraternity of futurists to ignore fatal flaws and defer to it as a landmark. Such, for example, has been Forrester's computerized simulation of a city.[1] Demonstrated here was the fact that only the most arbitrary assumptions, for example, unchanging environment, no suburbs, and external funding, could make the model "work." The procedures used by Forrester to compare alternative policies were, contrary to the systematic and methodological pretensions of the exercise, intuitive, policy being adjudged desirable or undesirable without elucidation as to the basis for such evaluations. In his comment on the role of computer simulation models in the design and testing of alternative urban policies, one reviewer of Forrester's work observed that

there are risks in the extension of "systems analysis" to social problems: it requires both extrapolation of inadequate behavioral theories and assumptions about subjective values. The impressive combination of confident technician and massive IBM computer must not be allowed to obscure those risks.[2]

THE STATE-OF-THE-ART OF FUTUROLOGY

In view of the limitations with respect to one simulated city, extension of the technique — in space, to include all cities, the nation, the world, or the universe; and in time, to encompass the rest of this century and part of the next — might seem ill-advised. And yet this is the determined activity of a number of organizations. One of them, the Club of Rome, is attempting to "simulate the reality of the world through mathematical insight." Funded by the Volkswagen Foundation, the group uses the Forrester model as its prototype. Its computer technologists having arbitrarily selected five main values as the ones important in the whole world, interlinkage is made by some eighty nonlinear equations "devised from the best information available in the international organization." The ultimate objective of this ambi-

[1] Jay W. Forrester, *Urban Dynamics,* Cambridge: MIT Press, 1969.
[2] James Hester, Jr., "Systems Analysis for Social Policies." Review of Jay W. Forrester's *Urban Dynamics, Science,* Vol. 168, No. 3932, May 8, 1970, p. 694.

tious undertaking is "to find ways to project which institutions and which processes will be necessary if we are to reach the point where there is some global management of the whole world."[3] Presumably, management is the *summum bonum*, and everyone on the face of the globe will live happily ever after in the computerized paradise engineered with benevolence aforethought.

The systems approach attacks the future in much the same way as it deals with the present. There is the same pseudo-serendipity that "discovers" paths long trodden; there are the same tools and techniques, used now, however, without the few constraints imposed by real-life tests in the present. There are the data banks, which, according to the futurist's handbook, must be appropriately stocked and organized and related to formal models of important dynamics. The fancy guesswork, with all its technological embellishment, of simulation and gaming, Delphi techniques and scenario construction, will serve to "invent credible paths between present conditions and hypothetical future states."[4] Embedded in this bill of particulars are many normative, methodological, and unsupported presuppositions. The assumptions are also made that information in data banks is, or can be, appropriate to needs not yet defined and organized according to specifications not yet delineated; and there are formal models both adequate and so future oriented that they anticipate what may at some later point in time prove to be important dynamics. As in the case of conventional applications of systems analysis techniques, where the more the critical observer knows of the specific field, the less convincing he is likely to find the "technically" contrived solution to its problems, so with the assessment of the art of the futurist, the farther away in time and the more widespread the uncertainty, the greater is the ease of acceptance of the grand plan.

Notwithstanding professionalization of the soothsayers' and seers' auguries in the form of games and other Delphian devices, or even R. Buckminster Fuller's "comprehensive anticipatory design science," the design of the future is little more than an image projection, more revealing of its creator's *Weltanschauung* than of the form and direction of social changes ahead. The model he devises, whether he knows it or not, epitomizes the basic philosophical conflict between free will and determinism. The will exercised is, of course, his, for he has made

[3] Aurelio Peccei (Managing Director, Italconsult), "Models and World Systems," Berkeley, California, University of California, Institute of International Studies, Guest Lecture Series, *Prospects for a Future "Whole World,"* April 1, 1971 (mimeo).
[4] Marvin Adelson, "The Technology of Forecasting and the Forecasting of Technology," Santa Monica: Systems Development Corporation, SP-3151/000/01, April, 1968, p. 19.

a number of important value-laden judgments and choices. Determinism is displayed in the very conception and operation of the model. Based on and extrapolated from a view, however eclectic, of the known present, it has certain "logical" and, therefore, unavoidable conclusions, which, moreover, are in the nature of the closed loop of a servomechanism. The whole process of systems analysis, or social engineering, takes on a decidedly architectonic thrust when applied to the future. Because of the likelihood of prediction feedback and the opportunity for advocacy, open or covert, of a particular course of action, the methodology provides the makings of a self-fulfilling prophecy. Popper reminds us of the influence that prediction may have on the predicted event and calls it the "Oedipus effect." [5]

Bertrand de Jouvenal describes the process as follows:

Any so-called "prediction" is always a starting point for examination of what *should be done on the assumption that it is true*, but always also *an outcome of assumptions concerning what will have to be done to make it come true*.[6] (My italics.)

Self-fulfillment is bound to come about when the essential components that are selected are organized in such fashion as to make the prediction come true.

Systems analysts and other futurists who have taken upon themselves the task of designing a better future seem to regard the undertaking as their private Promethean burden. Bauer, for example, describes the problem of foreseeing the future as among the most difficult and unsolved as any with which man is confronted. "Yet it is a problem which is inescapable," he avers without, however, supplying the reason for having assumed a task so thankless and hopeless. His exposition comes to the conclusion that a possible solution must be attempted "no matter how poor the result." [7] The assumptions built into the logic here and elsewhere in the futurists' approach are that the future can be "studied" or "foreseen"; that there are methods by which to predict social change; that social change should be planned and controlled, even though the techniques are admittedly crude and the results "poor"; that application of their "rational" procedures will guarantee a better future than some other, perhaps less "rationally" devised.

Unravelling the mystery of the future was once the bailiwick of sooth-

---

[5] Karl R. Popper, *The Poverty of Historicism*, London: Routledge & Kegan Paul, 1957, p. 13. In the legend, Oedipus killed his father, whom he had never seen; this was the direct result of the prophecy that had caused his father to abandon him.

[6] Bertrand de Jouvenal, "Notes on Social Forecasting," in *Forecasting and the Social Sciences*, Michael Young, ed., London: Heinemann, 1968, p. 120.

[7] Raymond A. Bauer, "Detection and Anticipation of Impact: The Nature of the Task," in *Social Indicators*, Raymond A. Bauer, ed., Cambridge, Massachusetts: MIT Press, p. 17.

sayers with omens to scrutinize, seers with crystal balls, and astrologers with their charts. The Cumaean Sybil and the Delphian Oracle of antiquity are the old mythology. The new mythology has developed its own shibboleths. The mystique of futurism purports to study the future scientifically, explore alternative futures rationally, and thus design the best of all possible futures. In assuming that that which has not yet happened can be studied, the futurists proceed without acknowledging their intellectual indebtedness. Actually, they follow the footsteps of philosophers of history, especially those historiographers who have tried to discern and analyze recurring patterns of the past as referents for the pattern of the future.

Oswald Spengler traced the decline of many civilizations and foresaw decay and ruin as inevitable.[8] Toynbee's hypothesis [9] was that civilizations crumbled when they failed to meet certain challenges. The ray of hope that would save twentieth century Western man was the salvation to be achieved through religious penitence. For Sorokin,[10] whose cultural dynamics were derived from a kind of Hegelian dialectic, religion itself was a manifestation of the prevailing social and cultural *Zeitgeist*, and neither extraneous to nor exerting influence on it. As a result of the inexorable "law of immanent causation," religion and all other manifestations of experience in the Sensate period, which Sorokin designated as ours, are sensual, secular, and non-transcendental.

That no one true pattern of social change has emerged from attempts at systematic study of the past is manifest in the internal inconsistencies within the divergent theories, the contradictions among the theorists, and the generally Procrustean treatment of intractable events of history resistant to the pre-set mold into which the historiographers tried to cast them.[11] Hindsight seems to suggest that the clue to predicting most accurately the shape of things to come lies in the rare and brilliant intuition that identifies a key dynamic aspect of the social order and perceives its potential developmental significance. James Bryce [12] for his time, Gunnar Myrdahl [13] for our time, and Alexis de Tocqueville [14] for all time exemplify the durability, if not permanent verity, of forecasting based on the combination of social insight, experience, and judgment. Their approach may not have been necessarily right, but because it made no pretensions to "rational," "scientific," or "logical" methodology, it

[8] Oswald Spengler, *Decline of the West*, New York: Knopf, 1929.
[9] Arnold J. Toynbee, *Civilization on Trial*, London: Oxford University Press, 1946.
[10] P. A. Sorokin, *Social and Cultural Dynamics*, Boston: Porter Sargent, 1957.
[11] Karl R. Popper, *The Poverty of Historicism*.
[12] James B. Bryce, *The American Commonwealth*, London and New York: Macmillan, 1888.
[13] Gunnar Myrdahl, *An American Dilemma*, New York: Harper and Brothers, 1944.
[14] Alexis de Tocqueville, *Democracy in America*, Henry Reeve text, revised by Phillips Bradley, New York: Alfred A. Knopf, 1963.

could be readily evaluated and assessed, accepted or refuted, whichever seemed reasonable.

Such is not the case with the social forecasting performed by today's futurists whose expertness has not been demonstrated convincingly with respect to understanding the past or present. Their scholarly conferences and compendia of papers having bravely thrashed the straw man of conventional statistics for being unreliable and inadequate, they try to develop "social indicators," which will somehow provide macro-insights into the multiplicity of changes still to come. Ignoring the perils of linear curve extrapolations, they repeat old mistakes by establishing their conception of a firm and reliable data base which then becomes their springboard into the future. Their literature stoutly affirms that the indicators need not necessarily be quantitative, and, in fact, that it is desirable that the qualitative be considered. But their focus and emphasis belie their heroic assertions. They start ambitiously with the universe or Planet Earth but soon whittle out a few variables which they and their computers can handle comfortably.

The mythology of systems analysis accompanies its forward march into the future. Presented as though it had accomplished wonders and taken the guesswork out of planning, the technique is represented as the key and clue to the salvation of mankind on this planet. Those who sell this notion believe their own sales story and they are finding buyers among decision-makers in the far flung corners of the earth. What is new and portentous here is that invocation of "scientific" tools and techniques, which provide a dutiful and convenient rationale for whatever cause of action seems politically expedient, may stifle thoughtful research and experimentation. Heady with heterogeneous facts and shy of theory, the futurists may be directly or indirectly abetting the anti-intellectualism that has already gained considerable momentum in this country and abroad.

Through propagandistic promotion, iteration, and reiteration, the tools borrowed from technology and the techniques derived from a heterogeny of disciplines are not precasting the shaping of the future. As the accepted means for controlling and directing social change, they have transformed futurism from the cynical game of the men under the Iron Mountain,[15] and others in "think tanks" secret and not-so-secret, into a game plan for the social order to come. The players are experts, entrepreneurs, and social engineering buffs who have persuaded themselves and a gullible public into believing that the future can be studied, that their methods should be used even though "crude" and leading to

[15] *Report from Iron Mountain on the Possibility and Desirability of Peace*, New York: Dial Press, 1967.

"poor results," [16] and that the future designed through "rational" procedures will be better than its much maligned, disorderly, democratic
alternative, arrived at through the presumed anarchy of social forces
without vector.

SYSTEMS ANALYSIS IN SOCIAL PERSPECTIVE

It is important that we concern ourselves with systems analysis as
it appears in the future tense; it is imperative that we keep our focus on
the self-fulfilling prophetic propensities implicit in its present usage.
We first must recognize systems analysis as more than an assemblage of
techniques and methods but rather as a social phenomenon fraught with
social significance, perhaps all the more because it is characterized by
contradictions, internal and external. Even though its assumptions and
presumptions lack empirical confirmation, it has, within a remarkably
short time, developed into a pervasive methodological ideology. The
state of its art paradoxically less advanced than that of many intellectual streams from which it has derived form but neither content nor
discipline, it has been accorded a prestige not earned and a respect not
demonstrably deserved. Its mentors the military and its models econo-
metric, its credibility has somehow managed to survive the defrocking of
McNamara and the plunge of his methods to the nadir of their popularity in the Pentagon as well as the disenchantment among economists over
preoccupation with the technique that has won international kudos for
their professional brethren. The very durability and resilience of the
systems approach is a factor worthy of note in a review of its
phenomenology.

Supposed to overcome the piece-meal fragmentation of other, more
specialized approaches, the systems approach has provided a language
that talks of total embrace of social processes and dynamics but delivers
methods that reduce wholes to their arbitrary and often least important
common denominators. Supposed to solve social problems, it has merely
served to redefine them in a way amenable to the technical treatment.
If, as we have observed in so many cases, an initial error was attributing
to certain categories of events more precision than was warranted by
their nature, then a cardinal sin was committed in the case of events
that were human and social. Experience has shown that the orderly and
predictable factors may, in the final analysis, be those of least importance
in the dynamics and direction of social change.

Carried to logical extremes, emphasis on quantification could so limit
and bias perspectives as either to distort and violate the essential nature
of social problems by forcing them into a tractable soluble state or to

[16] Bauer, *op.cit.*

institutionalize and legitimize neglect of them or their vital parts. All but forgotten in the methodological game-playing is the fact that the systems approach was supposed to encompass in its comprehensive grasp all facets and not a limited aspect of the matter under consideration. Merchandised as a Space Age specialty, a precise and sophisticated set of tools, systems analysis has become the stock-in-trade of practically any individual or organization seeking a government grant or contract or engaged in a project. Its language is the life line of everyone who aspires to make his work appear systematic or technically sophisticated. Deeper probe reveals how thin lies the veneer of glossolalia over fuzzy conceptualization and hyperkinetic data accumulation. With emphasis of both buyers and sellers of systems on quantity, qualifications are nebulous and quality control of output nonexistent.

As an instrument of public policy making, techniques of systems analysis have encouraged emphasis on the wrong questions and provided answers the more dangerous for having been achieved through a "scientific" or "rational" means. The ultimate result is a systematic foreclosing of promising avenues toward possible improvement and reform. Contrary to being an instrument of innovation, the systems approach is essentially reactionary. By defining problems in terms accessible to the tools, systems analysis has encouraged systematic neglect of facets and variables which could be crucial in both their generation and amelioration. In most social problems, even those attributable in large part to technology, aspects amenable to technical treatment are likely to be less important than those which are culture-bound, value-laden, and honeycombed with a political power network.

Cost-benefit ratios, program budgeting, and other procedures have forced preoccupation with only limited and arbitrarily delineated facets of public affairs, with the objective more likely to be bureaucratic self-justification than the general social welfare. Supposed to produce economies in cost of government and efficiency in operation, the technology of systems analysis with its hardware and software has burdened government decision-makers with elaborate mechanisms which use vast resources of money, time, and energy without demonstrably cutting costs or improving efficiency. Despite the exquisite calculating capabilities of computerized accounting, there has been no serious attempt at a cost-benefit study of systems analyses conducted at government expense. No one in government can tell how much is being spent on information systems and cost-benefit studies nor how much is involved in the frenetic nationwide switch to often unworkable and possibly already obsolescent program budgeting. And even if the costs could be enumerated, the benefits would be nebulous.

The application of the techniques may yet result in game-plan government relevant primarily to its simulated and skewed model but un-responsive to a fast-changing society. While administrators may have become peculiarly susceptible to and satisfied with symbolic solutions, the public at large is becoming more acutely aware of its rights and could demand more tangible evidence of concern for its welfare. Per-haps herein lie the seeds of a coming social revolution, one in which the technologically contrived and perfected image of a well-being never experienced will be the prime target.

Dazzled by the panoply of "Space Age tools" and overcome by the panegyrics of systems analysis enthusiasts who have made public prob-lem solving their business, administrators have been put on the defensive vis-à-vis their managerial efficiency. They have been captivated by the sophisticated techniques touted to be so potent elsewhere and try to im-prove performance, however ill-defined, by hiring outside experts as consultants. While use of outsiders to perform specialized tasks is not new, what is noteworthy here is the growing incidence of government-by-contract that removes from public officials responsibility of the deci-sions made. Because consultants are never held accountable for bad advice, the arrangement shields everyone from criticism. There is al-ready substantial evidence of dependence on the strategem of hiring outside experts to perform systems studies as a political ploy to convey the notion of official attention even when action is not politically feasi-ble or desirable. Moreover, because of the way in which systems analysis can be crafted to suit the occasion, the use of hired specialists may serve the politically useful purposes of masking bureaucratic ineptness and inadequacy, of providing support for a course of action already decided upon, or of working as a red-herring, diversionary tactic. Whatever else it accomplishes, the team of outside systems analysts, pre-empts func-tion and funds which might otherwise have enabled professional re-search. "Captive by contract" research blunts the edge of justifiable in-quiry and criticism and militates against exercise of intellectual au-tonomy that should be encouraged to make and keep government responsive to social and human needs. What is ominous is that there will always be willing mercenaries, some of them academic and all with an entrepreneurial bent interested in using closed-book, mission-directed analysis as a vehicle for personal fame and fortune. That this is happening at the very time that universities and institutions of higher learning are feeling the backlash of public disaffection may have signifi-cance the full dimensions of which are not immediately discernible.

The trend at all levels of government toward increasing involvement with private consultants has deeper implications than the sometimes mentioned threat to the Civil Service posed by circumvention of regula-tions through occasional hire of an outside specialist. Emerging as a factor in policy-making processes are the constituency of research institutes and corporations which, individually or in tandem, are becoming a kind of shadow government. Allocated contracts to execute the planning, design, implementing, and even evaluation of projects costing millions of dollars and thousands of lives, these techno-corporate entities have been seen as a force undermining the very form of government prescribed by the Constitution. Ready with a façade and made-to-order proposals designed to infuse confidence in their "systems capability," far-ranging consortia sometimes compete for and sometimes cooperate in transportation systems in the Northeast, low cost housing in St. Louis, rural development in Uruguay, and agricultural reform in Nigeria. Hired by and under the mission direction of regulatory agencies, consultive experts are in a position to perpetrate a kind of advocacy planning that has stunning potential for circumventing the checks and balances protective of the democratic process and influencing the shape and direction of domestic and foreign affairs. The selective examination and evaluation they perform lend the authority of technico-logical justification to regulations which may serve certain industries better than the commonweal. As adjuncts to advisory bodies, they are likely to be called on even more frequently in the future as public issues encompass considerations which are increasingly technological in nature. As they apply their skills in such situations, outside experts bring their own prejudices and predilections. In areas of social policy planning, consultants chosen for their "systems competence" could prescribe courses of action which could lead to a certain, but democratically unsafe and unsound, social order.

The game of musical chairs has long been played in government circles. Admirals and generals retire to become top potentates in industries that advise and sell their wares to the military. With every change in administration, the top echelons of appointees move out in a body. In recent years, many have joined research institutes, where they accept the commission to take on tasks which, as bureaucrats, they proclaimed were impossible. The advantage they distinctly enjoy in the new location is the freedom from the responsibility to provide workable, implementable, and realistic plans. As civil servants they had to fulfill certain obligations and meet some expectations; as freelance operators they have no such constraints and their output is subject not even to the modicum of quality control that may have been applied in the bureaucratic setting.

With the exodus to the consulting sidelines, there is the periodic influx of industry's best, armed with the latest tools and techniques for ensuring efficiency of operation. But, in the resulting blurring of lines between the public and private interest, the public's welfare receives low priority. The coalition of special interest and corporate power can influence decisions of far-reaching importance, with the pertinent considerations tailored to fit better the needs of certain select groups than the public at large. The role and function of government to protect the interests of the public become more and more attenuated as the key points in the decision-making apparatus are relinquished. Many regulations which should be enforced for the social good may be clearly uneconomical and not good business. But, when government accepts business as its model and economics as its decision-making means, and depends for guidelines on experts whose techniques have a strong bias, its goals are calculated with measuring devices of limited scope. In accepting and applying systems techniques, government planners have allowed the medium to become the message. The technique dictates the desiderata and assigns the priorities according to the pre-fabricated scenario.

While remarkably successful in becoming entrenched as the Space Age nostrum for society's ailments, systems analysis has not served as a likely cure for the aerospace industry's failing health. Having recognized its failure as a diversification tactic to supply job opportunities for the thousands of displaced workers, at least one representative has acknowledged that the systems management developed to such a high degree of sophistication in his business has little relevance in social concerns.[17] To suggest that systems analysis should not be hailed as the prime spinoff of the national space endeavor is not to belittle the accomplishments of the National Aeronautics and Space Administration nor any other bodies, public and private, that have engaged in the gigantic undertaking of getting men and instruments to the moon and beyond. Systems techniques, in myriad forms, may have played an important part in these activities. Certainly, the organizational skills displayed have been spectacular. But experience has shown that the methods are not universally applicable. Systems management, as applied in the Department of Defense and National Aeronautics and Space Administration, has little direct relevance in the social arena; social systems resist management. Social systems, as Lockheed's engineers discovered, are "complex, conflicting, and indefinable." [18] Their components cannot be treated like little black boxes and their goals are prismatic — a shifting mosaic of the society's values.

The variegated experts who have chosen to invade the market for systems studies and designs have been slower to acknowledge the inade-

[17] Dean S. Warren (Manager of Market Planning and Research for the Missiles System Division of Lockheed Missiles & Space Company), as quoted in "Humans vs. Hardware — A Critical Look at Aerospace as an Urban Problem Solver," *Aviation Week & Space Technology*, June 7, 1971, pp. 62–63.

[18] *Ibid.*

quacy of their tools for the job to be done. With technique their sole repertory, they ply their trade wherever there is a willing customer. And there still will be many. In the move into the systems field they have come from a heterogeny of intellectual backgrounds, in all of which certain bodies of theory and reservoirs of accumulated knowledge kept them within bounds. The farther from home base the experts have roved, the more attenuated has become the discipline. In fact, the less acquainted they were with the problem areas, the louder have been the pronouncements of the value of objectivity, the quality they were substituting for substantive knowledge. On this point, Popper's wisdom is especially cogent. He stresses the fact that scientific objectivity is not a product of the individual scientist's impartiality but rather an outcome of participation in a particular scholarly community based on the inculcation of standards of discourse and investigation as well as the public disclosure of methods and results.[19] So far as an individual scientist's objectivity can exist, it is not the source but the result of social or institutional arrangements governing the discipline. What passes, in the systems approach, for interdisciplinary effort turns out to be undisciplined activity, a game almost anyone can play. Nonetheless its advocacy assured through interest now vested, systems analysis and its surrounding methodology have become the prevailing style in public and private affairs. Even though fulfillment lags far behind iterated promise, proponents will entertain only those questions having to do with technical niceties.

Adherents of the approach usually counter critical review with an offensive: "What technique would *you* propose as an alternative to systems analysis?" "How would *you* improve systems to accomplish socially worthwhile purposes?" "To what better uses would you put systems analysis?" These questions, like the logic that prompts the technique, are inappropriate in the context in which they appear. The fact that they are raised at all indicates the extent to which the technological imperative enters into the phenomenology of the systems approach. They might appropriately be countered with, "Why should we use systems analysis in these matters at all?" "Why not explore means and methods better suited to the problem at hand instead of slavishly invoking techniques just because they are available?" Muddling through is probably safer in the long run than the wrong cure. Just because we have an arsenal of hydrogen bombs and powerful delivery systems need not mean that we search out a potential enemy to eradicate from the face of the earth. Similarly, we need not feel impelled to provide *ex post facto* justification through utilization of every technological device and development that springs from Aladdin's lamp. A technical

---

[19] Karl Popper, *The Open Society and Its Enemies*, pp. 405–406.

Chapter 9: *Futurology and the Future of Systems Analysis*

approach that may have served a useful purpose in one context may not be viable in another and may actually be detrimental.

The question we have asked in this research study is, "Are the techniques of systems analysis appropriate when we are dealing with problems which are essentially human and social?" The findings indicate that in their present condition they are not. And the direction in which they are developing promises little improvement. Refinement of methodology has led only to greater preoccupation with abstraction while the mythology that social problems can be solved remains unchallenged. In fact, this false assumption plays an important part in the perpetuation of the magic spell which promises a technology to solve social problems.

This is not to say that systematic approaches do not have a contribution to make to the understanding of social process and improvement of the social condition. The problems besetting mankind are plentiful, complex, and multi-faceted enough to provide challenge to and invite the commitment of professionals from a variety of disciplines. The clearly non-linear, normative, and value-laden dimensions of these problems need deter the efforts of only those experts who approach with predetermined solutions. The systems approach, if it is ever to become conceptually sound, must be a genuine multi-disciplined endeavor, in which contributions from the pertinent fields of knowledge are meaningfully synthesized, and not merely homogenized into a synthetic and symbolic language.

Based with some degree of confidence on the empirical evidence, the rebuttal to assertions of defensive support for current systems analysis as the answer to society's problems could state the known truth that, despite the methodological, systematic, and systemic pretensions of systems analysis and systems analysts, there is no single method for all problems for all people at all times. There is no cosmic scale solution. The appropriate approach is a function of the particular problem, the particular researcher, and the attendant circumstances. Each analyst must seek out, develop, and apply the particular set of tools required for the task at hand. The outcome of his work will probably not be perfect, but he will not feel called upon to rationalize his results or justify his course of action through manipulation of technicalities. Amendments and improvements will occur, if ever, on the real-life scene and not on the shadow screen reflecting the playing out of a scenario. To the oft-iterated counter argument that one should not criticize systems analysis unless one can supply something better, there is an answer — competent research and experimentation, with conceptualization first, technique last, and professional judgment always.

10. Beyond open system models of organization
Louis R. Pondy

Originally presented at the Annual Meeting of the Academy of Management, Kansas City, Missouri, August 12, 1976. The *E:CO* editorial team would like to thank the University of Illinois for their kind permission to print this article for the first time.

Introduction to Pondy's "Beyond open system models of organization"

I was in his office, when Louis R. Pondy read the letter rejecting his 1976 article at *Administrative Science Quarterly* by decision of the editor, calling him a member of the "cute school" of organization theory. The paper was revised, with Ian I. Mitroff (Pondy & Mitroff, 1979), and published in obscurity. Over the next decade Pondy's solo article, circulated in the academic underground, became widely recognized as a seminal piece of system theory, often citing the section: "Human organizations are level 8 phenomena, but our conceptual models (with minor exceptions) are fixated at level 4, and our formal models and data collection efforts are rooted a levels 1 and 2" (Pondy, 1976: 6).

Pondy's main contribution was to point out the role of language that was more than sign-representation, where image, story, and symbol, as well as self-reflection interpenetrate complexity. As a story theorist, I was influenced by Pondy's call to get to a 'languaged' understanding of storytelling in social organizations (level 8). My contribution is to illustrate the timeliness of the nine levels model by showing its story implications to complexity. In Table 1, in first four levels, it is narrative, not story that is dominant property. Narrative is a restricted type of story, that since Aristotle (350 BC) must have beginning, middle, and end, and be just a few episodes that imitate experience, and not a more epic-story (which includes simultaneous telling by many more characters and more imitative incidents and detail).

It is at level 5 (organic) that the more epic aspects of story begin to emerge, and the first of four Bakhtin (1973, 1981, 1990) dialogisms emerge (*polyphonic*, meaning many voices and logics in dialog in the tellings). Level 6 is where image stories emerge, in a multiplicity of modes (written, oral, architecturally, etc.), in a *stylistic* dialogism (a dialog among the stylistic modes where no one mode captures the image story). Level 7 the more *chronotopic* dialogism emerges as a property of system complexity.

Stories, narratives, and metaphors highlight certain aspects, and hide others from view. Level 2 (mechanistic) level 3 (control) in Table 1, objectify organizations, as does the organic metaphors (Boulding, e.g., uses cell for level 4; plant for level 5, and animal for level 6). As one moves up the levels of system complexity, the role of language is more pronounced (level 6 image; level 7 symbol; level 8 social organizations (or multi-cephalous, Pondy's substituted metaphor); level 9 transcendental).

The narrative prison gives way to many ways of storying temporality and spatiality, or what Bakhtin (1981) terms the relativity of space/time. At level 8 a more *architectonic* dialogism (defined as inter-animation of cognitive, ethical, and aesthetic discourses) emerges as a property of system complexity. This is the level at which societal discourse, as Boulding (1956) remarks impacts upon the social organization, where a given organization is playing its role begrudgingly in a 'network' of organizations and communities. Boulding (1956) and Pondy (1976) had no conception of the dialogisms, or how they might interplay at level 9. Pondy wanted to leave level 9 unspecified; Boulding, however, labeled it 'transcendental' and commented on the relation of what is unknowable, to what is posited as knowable. For me, this translates into a more spiritual awareness of how unknowable registers patterns in social behavior, and more explicitly in story, despite the triumph of modernity in making transcendental reflexivity in science taboo.

Pondy's work went unchallenged, until Robert Cooper (1989: 479-489, 495-500) critiqued it severely for challenging an input-output model of communication in system theorizing which Pondy did not escape even by using Chomsky. I answer Cooper's challenge above by pointing out the dialogisms. Cooper also claims Pondy's rendition of Boulding lacks the deconstructive cast Jacques Derrida brings to writing about system theory. Here too, I would say both Boulding and Pondy were doing moves that Bakhtin (1973, 1981) terms 'heteroglossic' which critics now argue presages Derrida's deconstruction theory. Specifically, heteroglossia is a language theory where centripetal (deviation-counteracting, such as level 3) language forces are being opposed by centrifugal (deviation-amplification) forces of language that commence at level 4 (open) and continue with levels of image, symbol, and through the four dialogisms (polyphonic, stylistic, chronotopic, and architectonic). I therefore claim Pondy's article remains seminal because it anticipates several of Derrida's

1. FRAMEWORK: Unique property is mapping the types of narratives in use by 'storytelling organization'. Narrative maps, such as tragic, romantic, comedic, satiric are dominant framework, but have forgotten 'epic-story' interactivity with such narrative types.
2. MECHANISTIC: Unique property is metrics to index time and travel motions of narratives throughout segments of the 'storytelling organization'. There are narratives about eras, summarizing everyday crisis points, but no biographical narration; time and space are mechanistic, imitative of machine metaphorization.
3. CONTROL: Unique property is centralized 1st-order cybernetic control of narratives behavior with rule-based mechanisms of deviation-counteraction. Observed narrative-scripts are compared to idealized narrative by a core of specialists noting deviations from the rules. 1st order cybernetics is a complication of machine metaphorization.
4. OPEN: Unique property is 2nd-order cybernetic deviation-amplification which opposes the level 3 deviation-counteraction, to accomplish self-maintenance while a throughput of narratives of the environment are sorted, without much or any knowledge generation. 2nd order cybernetics is further complication of machine metaphorization.
5. ORGANIC: Unique property is division of labor among mutually interdependent parts, each with highly specialized narrators, not doing much more than filtering environments for positive or negative narratives. Organization metaphorization framework, and machine-cybernetics coexist with mimicry of tree or plant. Up to this level narrative imprison story in sign re-presentations; mimicry of more polyphonic polylogic is emergent, not pronounced.
6. IMAGE: Unique property is the beginning of self-awareness and the narrating teleologically. Storying image of organization is differentiated from throughput processing of environmental narrative. Small antenarratives (bet and before fragments that aspire to narrative coherence) can transform a petrified image story. Metaphorization is computer-screen image mimicry achieved through orchestrating stylistic dialogism strategically.
7. SYMBOL: Unique is self-consciousness achieved through self-reflexive storytelling. Stories are self-reflexively co-produced and co-interpreted at symbolic level, not just mere signs or images. The 'chronotopicity' of constructing the time/space conception of history is salient. Metaphorization is history mimicry through chronotopic dialogism.
8. NETWORK: Unique property is self-reflexive awareness of organization situated in network of societal discourse that proscribes roles to organizations that organizations rescript. Architectonic inter-animation of cognitive, ethical, and aesthetic discourses, of which story is domain is manifest. Metaphorization is mimicry of computer network.
9. TRANSCENDENTAL: Unique is self-reflexive awareness of metaphysics of what is unknowable in opposition to what is knowable. Metaphorization is mimicry of spiritual enlightenment. However, the prior eight properties are still in interrelationship, including the Polypi dialogism of dialogisms of polyphonic, stylistic, chronotopic, and architectonic dialogism. Story has escaped narrative prison, but the narrative police are trying to arrest story (as always).

Table 1 *Nine levels of 'storytelling organization' complexity*

deconstruction moves. Pondy's reliance upon Chomsky lacks the dialogical explicitness of Bakhtin's philosophy of language.

Readers of system complexity may be more familiar with Fred and Merrelyn Emery's version of open system theory, that is rooted in Pepper's (1942) world hypotheses: level 1 (frameworks) would be formalism, levels 2 and 3 (mechanistic and control) would be mechanism; level 4 (open) would be contextualism; level 5 (organic) would be organicism. Merrelyn Emery (1997) says, "Open system theory is alive and well" and firmly rooted in Pepper's *contextualism* world hypothesis: open system and environment changes over time through series of historic events; and that the formist, mechanistic, and organic levels in Table 1, theorize closed and static systems. She asserts von Bertalanffy (general system theory) is an incomplete conception of *open system*, which was corrected mostly by Emery and Trist's (1965) laws of causal-texture mutually of organization and environment, and finished up by Ackoff and Emery (1972) with 'on purposeful' and 'ideal-seeking' laws of open system to become an ecological 'learning organization' (Emery, 1997). The point here is that for the Emerys, the 'organic' level of Boulding/Pondy would be less complex, than open system, and the Emerys's open system theory would contain aspects of image (level 6), but nothing about any higher levels.

Cooper (1989: 495) critiques system theory, in general, for being tautological with many empirical studies "mimicking" the subject matter, for concealing the exercise of power/knowledge, and domination in the way we work and live in organizations. For example, Katz and Kahn's (1966) system theory is said by Cooper (1989: 495, citing Degot 1982: 637) to be tautological: "the organization is a system ruled by the system laws: the variability identified in this system are created by laws whose form is such precisely because the organization is a system." The same critique would apply to the Emerys's 'open system' that the system is open because it follows laws of open systems. The problem is there is an unexamined open system naturalness (as organization and environment mutually adapt) that privileges essences (tautological laws), and puts the processes of system consummation beyond critical analysis. I counter, Boulding as well as Pondy address teleology as one of the levels in system modeling (level 6) where there is some self-awareness and teleological behavior, but that self-awareness is not self-reflexivity (a property key to identifying a level 7 system) has yet to emerge.

Pondy critiques Thompson (1967) who draws upon Katz and Kahn's reading of open system, which Pondy (1976: 10) argues is a level 3 system theory, that calls for the field to move beyond the closed system equivalent of Boulding's (level 2) clockwork. In short, neither Katz and Kahn (1966) nor Thompson (1967) is theorizing at level 4 open system; both apply Ashby's Law of Requisite Variety, which is a (level 3) control system logic (thermostat is Boulding's label for level 3).

The craft of consummating systems out of narrative that is at levels 1 to 4, quite monologic, and does not get at poly-logicality, an emergent property beginning in level 5 complexity, has yet to be addressed in theory or research.

While Cooper (1989) is accurate in pointing out that Derrida goes farther in developing a languaged theory of systems, than Pondy, call it light years ahead is storyteller's hyperbole.

David Boje

References

Ackoff, R. L. and Emery, F. E. (1972). *On purposeful systems: An interdisciplinary analysis of individual and social behavior as a system of purposeful events*, New York, NY: Aldine-Atherton, ISBN 0202307980 (2005).

Bakhtin, M. M. (1973). *Problems of Dostoevsky's poetics* , C. Emerson (ed./trans.), Manchester, UK: Manchester University Press, ISBN 0816612285 (1984).

Bakhtin, M. M. (1981). *The dialogic imagination: Four essays by M. M. Bakhtin*, M. Holquist (ed.), Austin, TX: University of Texas Press, ISBN 029271534X (1982).

Bakhtin, M. M. (1990). *Art and answerability*, M. Holquist (ed.), V. Liapunov (trans.), K. Brostrom (suppl. trans.), Austin, TX: University of Texas Press, ISBN 0292704127.

Boulding, K. E. (1956). "General systems theory: The skeleton of science," *General Systems*, 1: 11–17, 66–75. Also available in Richardson, K. A., Goldstein, J. A., Allen, P. M. and Snowden, D. (eds.) (2005). *E:CO Annual Volume 6*, Mansfield, MA: ISCE Publishing, ISBN 0976681404, pp. 252-264.

Cooper, R. (1989). "Modernism, postmodernism and organizational analysis 3: The contribution of Jacques Derrida," *Organization Studies*, ISSN 0170-8406, 10(4): 479-502.

Degot, V. (1982). "Le modele de l'agent et le probleme de la construction de l'objet dans les theories de l'entreprise," *Social Science Information / Information sur les Sciences Sociales*, ISSN 0539-0184, 21(4-5): 627-664.

Emery, M. (1997). "Open systems is alive and well," paper presented at Academy of Management Meetings, ODC division session.

Emery, F. and Trist, E. (1965). "The causal texture of organizational environments," *Human Relations*, ISSN 0018-7267, 18: 21-32.

Katz, D. and Kahn, R. (1966). *Social psychology of organizations*, Hoboken, NJ: Wiley, ISBN 0471023558 (1978).

Pepper, S. C. (1942). *World hypotheses: A study in evidence*, Berkeley, CA: University of California Press, ISBN 0520009940 (1961).

Pondy, L. R. (1976). "Beyond open system models of organization," working paper of the Organizational Behavior Group, Department of Business Administration, University of Illinois (Urbana), published in this issue.

Pondy, L. R. and Mitroff, I. I. (1979). "Beyond open system models of organization," in B. Staw (ed.), *Research in organizational behavior*, Vol. 1, Greenwich, CT: JAI, ISBN 0892320451, pp. 3-39.

Thompson, J. D. (1967). *Organizations in action*, New York, NY: McGraw-Hill, ISBN 0765809915 (2003).

Beyond open system models of organization - Louis R. Pondy

Introduction

I n a recent book, Haas and Drabek (1973) categorize organization research into eight perspectives or conceptual models - rational, classical, human relations, natural system, conflict, exchange, technological, and open system models - and add a ninth of their own, a stress-strain model. Their typology, like most typologies that have been developed to describe approaches to the study of organizations, is empirically derived from the work of practicing organization researchers. It is therefore potentially an excellent frame for representing the past and the present. But it severely limits invention of the future.

Inventing a future for organization theory is my intent in this paper. Doing so requires a framework of possible approaches that is more open-ended than that of Haas and Drabek. To maximize its creative contribution to the field, the framework should be drawn from some *other* field, yet still be relatable to organizational phenomena. And it should embody a relevant systematic basis for extending organizational research in fruitful directions.

Kenneth Boulding's (1968) hierarchy of system complexity is a good candidate for such a framework and I shall use it to organize my ideas in this paper. He identifies nine levels of system complexity. The systems in question can be either 'real' systems (e.g., a cell, a chemical reaction, a tree, a bird, a man, a family). Or they can be *models* of those systems. But models are just idea-systems, so Boulding's hierarchy can be taken as a description of the complexity of *either* phenomena or models for analyzing those phenomena. This dual use of the hierarchy to describe both organizations and models of organizations will be helpful in clarifying the state and possible directions of organization theory.

It should be emphasized that adjacent levels in the hierarchy differ in complexity not merely in their degree of diversity of variability, but in the appearance of wholly new system properties. For example, the difference between open system models of level four and 'blueprinted growth' models of level five is the presence of the capacity for genotypic growth and reproduction.

Level 9: SYSTEMS OF UNSPECIFIED COMPLEXITY

Level 8: MULTI-CEPHALOUS SYSTEMS
    Social organization

Level 7: SYMBOL PROCESSING SYSTEMS
    Self-conscious language users (humans)
    Knows *and* knows that he knows

Level 6: INTERNAL IMAGE SYSTEMS
    Animals - specialized information receptors
    "Brain": receipt of information, leading to an 'image'

Level 5: BLUEPRINTED GROWTH SYSTEMS
    Plants - division of labor among cells

Level 4: OPEN SYSTEMS
    Self-maintaining, self-reproducing

Level 3: CONTROL SYSTEMS
    Thermostats: transmissions and interpretations
    of information

Level 2: CLOCKWORKS
    Simple machines, theoretical strcuture of
    physics, chemistry and economics

Level 1: FRAMEWORKS
    Geography and anatomy of the universe

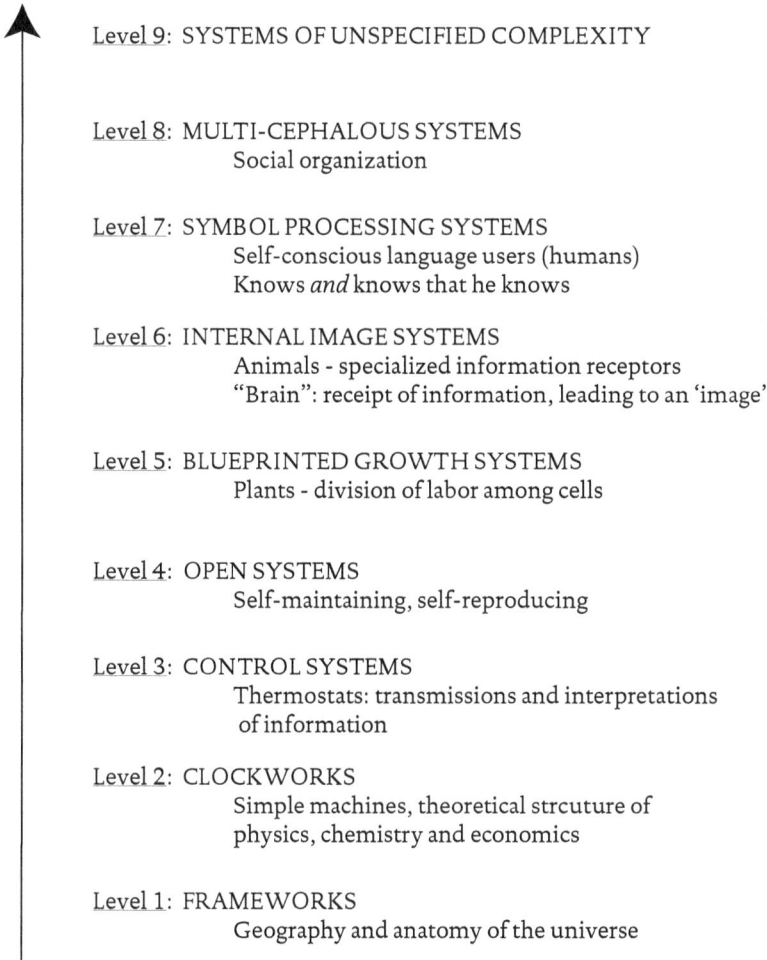

Figure 1 Boulding's hierarchy of system complexity

Boulding's hierarchy of system complexity
Level 1: Frameworks - Only static, structural properties are represented in framework models, as in descriptions of the human anatomy, the cataloguing system used in the Library of Congress, or an organization chart of the US Government. The latter may be complicated, but it is not 'complex' in Boulding's sense.

Level 2: Clockworks - Non-contingent dynamic properties are represented in clockwork models, as in descriptions of a precessing gyroscope, the diffusion of innovations, or economic cycles in a *laissez-faire* economy. The crucial difference from level 1 is that the state of the system changes over time. At any given time, level 2 phenomena can be described using a level 1 model.

Level 3: Control Systems - Control system models describe regulation of system behavior according to an externally prescribed target or criterion, as in heat-seeking missiles, thermostats, economic cycles in centrally controlled economies, and the physiological process of homeostasis. The crucial difference from level 2 is the flow of information within the system between its 'regulator' and its 'operator', and in fact the functional differentiation between operation and regulation. For a *given* control criterion, level 3 systems behave like level 2 systems.

Level 4: Open Systems - Whereas a control system tends toward the equilibrium target provided to it and therefore produces uniformity, an open system maintains its internal differentiation (resists uniformity) by "sucking orderliness from its environment" (Schrödinger, 1968: 146). Some people have mistakenly characterized an open system as having the capacity for self-maintenance *despite* the presence of throughput from the environment, and therefore have recommended buffering the organization against environmental complexity. Quite to the contrary, it is precisely the throughput of non-uniformity that preserves the differential structure of an open system. In an open system, the Law of Limited Variety operates: A system will exhibit no more variety than the variety to which it has been exposed in its environment. Examples of phenomena describable by open system models are *flames* (simple physical systems in which the transformation of oxygen and, say, methane into water, carbon dioxide and heat maintain the system's shape, size, and color), and *cells* (biological systems involving more complex transformations and more differentiated structures, and also the phenomenon of mitosis - duplication through cell division).

Level 5: Blueprinted-growth systems - Level 5 systems do not reproduce through a process of duplication, but by producing 'seeds' or 'eggs' containing pre-programmed instructions for development, as in the acorn-oak system or egg-chicken system. While the phenomenon of reproduction is not involved in language usage, the Chomskian distinction between the 'deep-structure' and 'surface-structure' of grammar seems to tap the same relationship as in acorn-and-oak. Both involve a rule-based generative mechanism that characterizes level 5 models. Explaining level 5 systems means discovering the generating mechanisms that produce the observed behavior. And *models* of level 5 systems will exhibit this dual level structure as well. (I shall interchangeably refer to models and systems at each of the levels, hopefully without confusion. The intent is that at a given level there is a structural isomorphism between the model and the system. Level 6 systems do, however, have level 5 properties that can be described using level 5 models, so that a system and a model of that system need not be at the same level. Thus open system models of human organizations can be built even though organizations exhibit higher levels of system complexity than level 4.

Level 6: Internal image systems - Level 3, 4, and 5 models incorporate only primitive mechanisms for absorbing and processing information. To quote Boulding, "it is doubtful whether a tree (level 5) can distinguish much more than light from dark, long days from short, cold from hot." The essential characteristic of level 6 systems is a *detailed* awareness of the environment acquired through differentiated information receptors and organized into a knowledge structure or image. (Boulding argues that his hierarchy is cumulative - each level incorporates all the properties of all lower levels. However, one might argue that some sophisticated computer software systems are at level 6, yet do not exhibit the blueprinted growth of level 5, unless one wanted to describe the relationship of programming languages to machine language as 'blueprinting'.) Level 6 systems do not exhibit the property of *self*-consciousness. They do not know that they know. That enters at level 7. A pigeon in a Skinner box and an organization that forgot why it instituted a certain rule might be examples of level 6 systems.

Level 7: Symbol processing systems - At level 6, the system is able to process information in the form of differences in the environment. But it is unable to generalize or abstract that information into ideas, and symbols that stand for them. To do that, the system has to be conscious of itself, and this is the defining characteristic of a level 7 system. It has to be able to form the concept '*my* image of the environment', and work on it. And to work on that image, it needs a coding scheme or language. So level 7 systems are self-conscious language users, like individual human beings. But it is less obvious that human groups are level 7 systems. I am not sure what it means for a group to have an image of its environment, unless the process of socially constructing a reality (Berger & Luckman, 1966) gets at it. And what does it mean for a group to be a language user as distinct from its members being so? Suppose the members all speak *different* languages. Then the group is not a language user, even though its members are, and it cannot construct a reality socially. But is a group a language user if its members *do* speak the same language?

Level 8: Multi-cephalous systems - Literally systems with several brains. Boulding's term for this level is 'social organization'. But the unit of analysis, the 'individual', is open to choice, and it might be confusing to refer to a social organization of persons versus a social organization of organizations. What is at issue is that the collection or assemblage of 'individuals', whether they be genes, or humans, or computers, creates a sense of social order, a shared culture, a history and a future, a value system - human civilization in all its richness and complexity, as an example. What distinguishes level 8 from level 7 is the elaborate shared systems of meaning that entire cultures, and some organizations (but no individual human beings) seem to have.

Level 9 - To avoid premature closure, Boulding adds on a ninth, open level to reflect the possibility that some new level of system complexity not yet imagined might emerge.

Having sketched out some features of Boulding's hierarchy of complexity, let me make a rash statement that I will attempt to justify. Human organizations are level 8 phenomena, but our conceptual models (with minor exceptions) are fixated at level 4, and our formal models and data collection efforts are rooted at levels 1 and 2. My worst expectation is that the field of organization theory will take its task for the next decade the refinement of analysis at levels 1 through 4. My greatest hope is that we will make an effort at moving up one or two levels in our modeling (both conceptual and formal) and begin to look at, for example, phenomena of organizational birth and reproduction, the use of language, the creation of meaning, the development of organizational cultures, and other phenomena associated with system complexity in the upper half of Boulding's hierarchy.

Since open system models have played such a central role in organization theory in the recent past, it would be useful to sketch the present view and some of the motives for change. Empty categories in a conceptual framework of approaches tend to suck a field in their direction, but they are insufficient to divert a field entirely from a useful paradigm. We must also show why the open system model, as it has been interpreted, is too limiting.

Open-system modeling
Open-system models have for the last decade dominated thinking and research in the field of organization theory. Most people would agree, I think, that Thompson's *Organizations in Action* (1967) comes as close as any treatment to being accepted as a paradigm statement

of the open-system perspective of what some might call the 'macro' level of organization. Actually, Thompson intended his book to be a reconciliation of the rational or closed-system model of organizations with the natural or open-system model, and his success in transforming a zero-sum game into a positive-sum game for the profession probably accounts for the enthusiasm that greeted publication of the book. Despite its reconciliatory intent, the book is dominated by an open-system perspective. (After all, *closing* the system is a meaningful act only within an open-system model!)

About the same time or slightly earlier, others besides Thompson (e.g., Lawrence & Lorsch, 1967; Perrow, 1967; Crozier, 1964; Burns & Stalker, 1961; Cyert & March, 1963) made important contributions to articulating the point of view that has subsequently permitted us to analyze and understand the problematic nature of uncertainty for the organization, and how uncertainty ties together technology, structure, and environment in a contingent relationship. The resulting paradigm statement generated a large amount of research and continues to do so. (For example, fully 40% of the articles in the 1975 ASQ cite *Organizations in Action*. By contrast, only 5% cite Weick's *The Social Psychology of Organizing*, a fact of relevance to my later discussion.) We have made substantial progress from where we were in 1967 in the direction pointed by Thompson. And despite Pfeffer's (1976) recent complaint that organizational behavior has been "dominated by a concern for the management of people *within* organizations," the 'organization theory' branch of organizational science *has* researched the organization-environment interface under the guidance of open-system thinking. So what's the problem?

Problem is, models not only direct attention *to* some phenomena and variables, but also *away* from others. And if a model is highly successful in helping a researcher to cope with problems the model says are important, habituation will take place: the researcher will simply not 'see' other problems, and he will have no basis for being receptive to competing models. But there *are* other problems that we should be addressing, and there are competing models that we should be considering. This is the motivation for my arguing that we need to go beyond open-system theory. Specifically, I offer five major reasons in support of this position:

1. By focusing on maintenance of the organization's own internal structure, open-system theory has directed us away *from ecological effects* - broadly defined - of the organization's actions, to the ultimate detriment of the organization itself.

2. We should be directing our efforts to understanding massive dysfunctions at the macro level, not just explaining order and congruence. How do organizations go wrong?

3. We need to reflect in our own models changing conceptions of man in other fields, especially those that increasingly picture man as having the capacities for *self-awareness* and the *use of language*.

4. Troublesome theoretical questions ignored by open-system theory are suggested by other models. For example, do organizations reproduce themselves? If so, how?

5. For the purpose of maintaining OT's adaptability as an inquiring system (albeit a loosely-coupled one), we need to discredit what we know, to change for the naked sake of change to prevent ossification of our ideas.

These motives for change are discussed more fully below. Following that some alternative models of organization are proposed. The paper concludes with a brief examination of the implications of my position for the doing and teaching of organizational research, and the teaching of present and future managers.

Motives for change: The limits of open system models
The ecology of organizational action

In order to understand how open system models can blind us to the nest-fouling impact of organizations' actions on their environment, we need to examine how open system theory has been interpreted and used by organization theorists. Frequently, those who claim to be using an open-system strategy are in reality using level 3 control system models. They have failed to make the distinction between 'natural' and 'open' system models (Haas & Drabek, 1973).
Consider Thompson (1967):

"Central to the natural-system approach is the concept of homeostasis, or self-stabilization, which spontaneously, or naturally, governs the necessary relationships among parts and activities and thereby keeps the system viable in the face of disturbances stemming from the environment" (p. 7).

In other words, the environment is a source of disturbance to be adapted to, instead of the source of 'information' that makes internal organization possible. *Self-stabilization* referred to by Thompson is a level 3 process. The equivalent level 4 process is *self-organization*. Haas and Drabek (1973) recognize the difference between natural and open system models, but classify Thompson incorrectly as an open system theorist! What Thompson calls a closed system is equivalent to Boulding's clockwork (level 2). Thompson made a major contribution by formalizing organization theory at a higher level. But it was *not* at the level of open systems! There is therefore considerable doubt whether organization theory (as represented by Thompson's book) is even *at* the open system level, to say nothing of whether it is ready to go beyond it. So this section will have to be split into two parts: (i) the ecological consequences of using a control system model (even though it is spuriously labeled as an open system model); and (ii) the ecological consequences of using a true open system model. By 'ecology' here I mean the structure of the organization's social, economic, and political environment as well as of its physical environment.

Control system thinking
We must remember that the aim of a control system is to produce uniformity, if it can. To the extent that the system environment is highly varied in its texture and over time, the regulator part of the system must match the variety of the environment so that it can control that variety and produce a uniform environment for its operator part. In Thompson's language, this means creating the conditions necessary for rational operation at the technical core. This is the essence of Ashby's (1956) Law of Requisite Variety. (It is not usually recognized that Ashby's Law is a statement about level 3 properties of systems, not level 4 properties.)

The ecological implications of control system thinking, both theoretical and practical, is that environments as well as organizations will become more uniform. Environments are made up of other organizations each of whom, according to this view, is following a control system strategy. Each attempts to impose uniformity on the others so that uniformity can be created 'inside'. The result is that the entire system will grind toward a social-system-wide equilibrium. Within the context of a control system model this is a desirable state of affairs. Not so for open systems.

Open system thinking
The ecological consequences of open system thinking are quite different. An open system is at such a level of complexity that it can maintain that complexity *only* in the presence of throughput from a differentiated environment. If an open system milks its environment of all its diversity and differentiation, then it will have only a uniform, grey soup to feed on, and eventually its own internal structure will

deteriorate to the point that open system properties can no longer be maintained. If control system models are used to 'manage' open systems, the system will be led to take precisely the *wrong* actions! The organization will attempt to insulate itself from the very diversity that it needs.

But suppose an open system 'realizes' that it needs its environment and does not attempt to buffer out variability. If environments are plentiful, and the system is mobile, it may still extract the needed organizing information from the immediate and present environment, leave it depleted (i.e., undifferentiated) and move on to another. But suppose environments are scarce. A system must then in some sense replenish its environment. It must, paradoxically, put variety back into the environment so that it can subsequently use it. But how to return variety to the environment without de-organizing the system itself?

The key to resolving this dilemma is realizing that only part of an organization's environment is *given* to it. Another part is *enacted* (Weick, 1969) by the organization itself. Some people have misunderstood Weick's concept of enactment to be identical with imagination or mental invention. But Weick means that the organization literally *does* something, and once done, that something becomes part of the environment that the system can draw on to maintain its own internal order.

There is a trap here to be avoided. If the enactments are merely an expression of the system's current organization, then nothing new will be created for the system to feed on. Complex systems have an appetite for novelty. They need what Stafford Beer (1964) has called "completion from without." Somehow, the process of enacting an environment must escape this redundancy trap. Weick (1976) has suggested a number of strategies that apply here: (a) be playful, (b) act randomly, (c) doubt what you believe and believe what you doubt (i.e., discredit the existing organization). All these strategies have promise of escaping the trap. Since there is a premium on this kind of inventive behavior when environments are scarce, we should expect to see playful, random, and discrediting behavior positively correlated with environmental scarcity (assuming that only the successful systems survive). If breaking the law can be thought of as discrediting, then this conjecture is consistent with the recent finding that violation of anti-trust regulations by corporations tends to increase when profits decrease (Staw & Szwajkowski, 1975).

Thinking of open systems as *needing* environmental variety also sheds fresh light on the widely replicated finding that organizational complexity is positively correlated with environmental diversity. The usual explanation from contingency theory is that the organization needs to be complex in order to cope with environmental variety. Implicit in this explanation is that 'surplus' complexity is possible but not necessary. The alternative explanation flowing from a proper analysis of open systems is that an organization is unable to maintain internal complexity except in the presence of environmental diversity. Surplus complexity is simply not possible from this view, but a shortage is. This might provide a basis for choosing between contingency theory and open system theory.

Hans Hoffman's view of the nature of man nicely captures this property of open systems, especially as it relates to the necessary character of the enactment process:

"The unique function of man is to live in close, creative touch with chaos, and thereby experience the birth of order" (quoted in Leavitt & Pondy, 1964: 58)

So, I have argued that organizations as open systems foul their environmental nests either by:

a. following a control system strategy and deliberately killing variety in the environment;
b. following a short-sighted open system strategy and failing to renew the successive environments that they occupy.

Open system theory as it is currently interpreted and practiced in the field of 'organization theory' does not come to grips with either of these problems. Important exceptions exist (Weick, 1969; Hedberg, *et al.*, 1976; Cohen & March, 1974), but they do not yet occupy center stage. (Recall my earlier comment that Thompson's book was eight times as frequently cited as Weick's in the 1975 *Administrative Science Quarterly*.)

Dysfunctions in organization theory
One of the striking differences between organizational behavior and organization theory is that OB defines much of its research effort in terms of dysfunctions of the system. For example, there are theories of absenteeism, turnover, low productivity, industrial sabotage, work dissatisfaction, employee theft, interpersonal conflict, resistance to change, and failures of communication. Even equity theory is really a theory of *in*equities, how they're perceived and how they're resolved.

But OT is a theory of order. We have theories of the *proper match* between structure and technology (Perrow, 1967), between environment and structure (Lawrence & Lorsch, 1967), between forms of involvement and forms of control (Etzioni, 1961). Thompson (1967) most eloquently speaks of administration as the 'co-alignment' of goals, technology, structure and environment, and he treats dysfunctions as neither serious nor permanent. Corrective mechanisms, true to the control system model, take care of any problems:

"Dysfunctions are conceivable, but it is assumed that an offending part will adjust to produce a net positive contribution or be disengaged, or else the system will degenerate" (Thompson, 1967: 6-7).

The prevailing view in OT offers no systematic typology of dyfunctions at the macro level. But the popular literature documents some of the more spectacular dysfunctions. For example, Halberstam (1972) has described the pressures for consensus decision making that operated within the Johnson White House to systematically exclude opposing points of view on our involvement in Viet Nam. Janis (1972) has done the same for the decision making that led up to the Bay of Pigs invasion in 1961. Smith (1963) described a number of crises in corporate decision making. And the more recent crises of bribery within Lockheed and the misuse of power within Watergate are familiar to the point of contempt. But organization theory as a field is so preoccupied with explaining order that it has not yet discovered these most interesting phenomena. (Note that it is not necessary to argue for the study of dysfunctions on normative grounds. From a purely descriptive, non-normative, perspective, such dysfunctions are intriguing scientific happenings.)

Consider Lordstown. Much has been written analyzing how and why the workers reacted to the speedup of the assembly line. But virtually nothing has been written explaining why General Motors made the wrong decision in the first place.

It's curious that in economics the situation is reversed. *Macro*-economics is focused heavily on the system dysfunctions of inflation, unemployment, and recession. But *micro*-economics is concerned with explaining the rationality of choice. Whether that reversal is significant I don't know. But in the organizational sciences, it's the macro branch that eschews the inquiry into disorder.

Like all attempts at generalization, this one suffers its exceptions. I have already mentioned Staw and Szwajkowski's (1975) study of anti-trust violations. And Staw (1976) and Staw and Fox

(1977) have studied the phenomenon of escalation (á la Viet Nam) experimentally. The use of power to influence the allocation of resources away from rational norms has been studied by, among other, Pfeffer and Salancik (1974) and Salancik and Pfeffer (1974). But the dominant thrust of the field has been explaining why organizations work well and do good.

Part of the responsibility for this harmony orientation can be assigned, I believe, to Thompson's use of *organizational* rationality as a central and integrating concept. By treating the unit of analysis as the organization plus the environment, we would instead be forced to define the bounds of rationality to be broader, to invoke a concept of *ecological* rationality (Bateson, 1972). It is not merely the organization that adapts to the environment. The organization and its environment adapt together. Within such a model of ecological rationality, the environment's problems become also the organization's, and dysfunctions come to be recognized as real phenomena worth explaining.

In a recent review of Schön's (1971) *Beyond the Stable State*, Rose Goldsen (1975) summarizes Schön's argument that institutional dysfunctions arise from a belief in the possibility of a stable state buffered against change and uncertainty:

"*Change and paradox are not anomalies to be corrected, but the very nature of open systems. Learning systems accept these principles as axioms, rejecting the 'myth of the stable state'. Our current institutions still base themselves on that myth and it is their compulsive insistence on trying to achieve it that leads to many dysfunctions and ultimate breakdown*" (Goldsen, 1975: 464).

Goldsen refers to Schön's redefining hotel chains as "total recreational systems," and then asks:

"*Is it dysfunctional when informational breakdown in the Coca Cola Company (say) interferes with efforts to maintain sugar economies in developing nations? Is it functional when 'recreational systems' convert large proportions of the indigenous labor force into waiters, bellboys and cab drivers, chambermaids and prostitutes? One man's 'dysfunction' is another man's 'function'*" (Goldsen, 1975: 468).

Within the paradigm represented by Thompson's seminal book, the effects described in the above quote would not be recognized. But Thompson's book was written ten years ago. The image of the world that it projects is a history of growth and prosperity, of munificent environments. But times have changed. It is no longer an accurate description of the world we live in. Nor is it a sensible guide to solving the problems our institutions face and have created.

If studying the conditions of order is incomplete, how shall we change what we do? I would argue that we need to develop a theory of error, pathology, and dis-equilibrium in organization. And open system models as currently interpreted are of little help for that purpose.

Alternative conceptions of man and method

A third reason for needing to go beyond the open system model of organization is that it excludes many fruitful models of human behavior. Organization theorists seem to have forgotten that they are dealing with *human* organizations, not merely disembodied structures in which individuals play either the role of "in-place metering devices" (Pondy & Boje, 1976) designed to register various abstract organizational properties (e.g., complexity, formalization, etc.), or the role of passive carriers of cultural values and skills. Thompson's conception of the individual is that society provides a variety of standardized models of individuals that organizations can use as inputs:

"...if the modern society is to be viable it must sort individuals into occupational categories; equip them with relevant aspirations, beliefs, and standards; and channel them to relevant sectors of 'the' labor market" (Thompson, 1967: 105).

Following Thompson, the vast majority of organization theorists have downplayed man's higher capacities, including his ability to use language, his awareness of his own awareness, and his capacity to attribute meaning to events, to make sense of things. These capacities are characteristic of Boulding's level 5 through level 8. They are also characteristic of that property called mind; and what we need to do is to "bring mind back in" (Pondy & Boje, 1976) to organization theory. A small number of organization theorists have made language, awareness, and meaning central concepts in their theories (Weick, 1969; Silverman, 1971), but the dominant trend is still toward mind-less conceptions of organization.

Not so with some other social science disciplines. Consider cultural anthropology. Geertz (1973) in *The Interpretation of Cultures* starts out by assuming that assigning meaning to events is a central human process, and that the task of the anthropologist is to ferret out those meanings and the meanings that lie beneath them in multiple layers. To describe only the events is 'thin' description, but to describe the layers of meaning underlying those events is 'thick description'.

One important class of meanings is the set of beliefs about causality. To Geertz these would be problematic, requiring explanation. But to Thompson, they are given in society and organization members are simply 'equipped' with them. But how do those beliefs originate and change with experience? Perrow (1967) has been influential in getting us to think of technology as well or poorly understood, as though technology could be understood without someone to do the understanding. But since technical knowledge varies from individual to individual, the degree of understanding is clearly a property of the object-observer pair, not of the object alone. Similarly, environments are not uncertain. (We do not even deal here with the more serious problem of how an organization decides where it leaves off and the environment begins. That boundary, too, is problematic (Weick, 1976). We have been describing environment and technology 'thinly'. A thick description would probe into environment and technology as ways of classifying experience and thereby giving meaning to it.

Consider Harre and Secord's (1973) recent reconstruction of social psychology. They propose an "anthropomorphic model of man," in which man is treated, for scientific purposes, as if he were a human being! That is, man is endowed not only with an awareness of external events, but an awareness of his own awareness (Boulding's level 7) and with a capacity for language. Most importantly, man is presumed to have generative mechanisms that produce observable behavior. The task of inquiry is to discover those mechanisms for each individual. In the prevailing OT paradigm, no such mechanisms are presumed. The form of explanation is therefore necessarily comparative across organizations at the same level of abstraction. But with a presumption of a "deep structure" (Chomsky, 1972) that generates the "surface structure" of observable behavior, a "theory of the individual" (Newell & Simon, 1972) makes sense, and a "science of the singular" (Hamilton, 1976) based on a case study methodology becomes rigorous science. The performance programs proposed by March and Simon (1958) are precisely such 'generative mechanisms' that produce organizational behavior. It is curious that organization theory should have drawn so heavily from parts of March and Simon, but missed that central point of the book for the most part.

The existence of alternative models of human behavior is insufficient by itself to cause us to desert open system models. But these alternative conceptions expose phenomena that the prevailing view cannot begin to handle. In short, the higher mental capacities studied by other disciplines offers a new avenue for OT to explore to gain fresh insights into organizational phenomena. With isolated exceptions, those new opportunities have not been explored.

New theoretical questions

I have previously argued that Thompson's open system model is inadequate for dealing with important practical problems. But it is also inadequate for conceptualizing some important theoretical questions as well. No theory should be expected to cope with the full range of phenomena, of course, but neither should allegiance to a theoretical position become so strong that it prevents us from considering phenomena outside its purview.

One important class of theoretical questions addresses the phenomena of organizational birth and reproduction. Extant open system models, for the most part, are about *mature* organizations. Although Thompson discusses some aspects of growth, his analysis is about continued growth of adult organizations. And it is growth whose patterns are shaped by external forces, not the blueprinted growth of Boulding's level 5. The same is true of the best known treatment of organizational growth and development (Starbuck, 1971).

Biological analogies can sometimes be carried too far, but in this case I believe it is useful to ask whether organizations 'reproduce' themselves in any sense. Consider the following model.

1. Organizational development is constrained by environmental forces, but it is directed by fundamental rules for organizing stored inside the organization itself. Those organizing rules, or generative mechanisms, produce the observed patterns of differential functioning that make up the organization.

2. The organizing rules are stored in the brains of some, perhaps all, individuals in the organization. But those rules result from a previous process of negotiating the organizational order.

3. When a person leaves the organization, he carries with him those organizing rules. Should he be the founder of a new organization, those rules would find expression through unfolding in a new environment.

This is essentially the underlying model in a recent analysis by Kimberly (1976) of the birth of a new medical school at the University of Illinois. At first glance, Pettigrew's (1976) analysis of entrepreneurship seems to tap the same phenomenon, but I believe something distinct is at work in Pettigrew's model. Whereas Kimberly is conceptualizing organizational birth as a reproductive process through the mechanism of 'offspring', Pettigrew seems to have a model of autonomous birth in mind. The entrepreneurs whom he has studied have formed organizations on the foundation of creative, novel myths

212   Classic Complexity pp. 193-226

or cultures. Those entrepreneurs don't seem to have come from any previous organizational experience.

A second important class of theoretical questions outside of Thompson's model is that dealing with higher mental capacities. I have already alluded to some of the work in the area. Just a few more comments here: How people make sense of their experiences is a crucial issue for organization theory because the answer potentially overturns models of rational behavior. A phenomenological approach is that sense-making is retrospective; we can understand what we're doing only after we've done it. An action-theoretic approach argues that meanings are socially constructed, that therefore there are multiple realities. These positions have been most systematically developed by Weick (1969) and Silverman (1971), but their influence on empirical research on organizations has been minimal. Weick's and Silverman's models are extreme points within organization theory, especially for American organization theory. But the view they advance is more accepted within anthropology, recent advance within the philosophy of mind, and European organization theory.

It is not immediately obvious why Thompson's model precludes such theoretical questions. As I have suggested, part of the reason is that Thompson seems to have mature, already organized systems in mind. What is problematic for him is simply maintaining that organization, not creating it in the first place. A second source of blockage is the causal priority Thompson assigns to norms of rationality. How the organization comes to articulate these norms of rationality is not problematic in Thompson's model, except to say that the organization goals are negotiated within the dominant coalition. But each member of the dominant coalition is presumed to have specific interests already in mind. An alternative model reduces rationality norms to retrospective outcomes. We have not yet discussed the role that language plays in this rationalizing process, but work from other fields suggests that terms in our language affect what we see (Whorf, 1956) and even the logic we use to structure our thought (Tung-Sun, 1970; Alexander, 1967: 39-47).

Change for change's sake

As my colleague Michael Moch has impressed on me, OT is not only a field of study, it is itself "an organization of organization studiers," albeit a loosely-coupled one. Since we only meet infrequently, it seems appropriate to refer to ourselves as an *occasional organization*.

If face-to-face interaction is a criterion for a manifested organization, then OT (or any dispersed scholarly field) spends most of its time underground, surfacing only occasionally during conventions. Although occasional organizations are loosely-coupled through personal interaction, they can be tightly coupled through common adherence to a system of belief, as in the case of a dispersed scientific community's adherence to a paradigm. Such a paradigm is used not only to guide research by full members of the community, but is also the basis for indoctrination of new members.

If the paradigm is too well-defined, or is believed in too strongly, then creative ideas inconsistent with the paradigm will gradually be selected out. If the field is to continue to be effective in working on worthwhile problems, then it must, to a certain degree, discredit what it knows and act hypocritically (Weick, 1976).

We need to maintain a certain creative tension in what we take to be true. Our system of scientific beliefs should be a *nearly* organized system - organized enough to provide the confidence for researching uncertain topics, but not so organized that doubt is no longer possible. The illusion of success, especially when it's hard-won, breeds resistance to change. Scott (1976) has voiced a similar concern:

"After searching so long for 'the one best way to organize,' this insight [contingency theory] *was hard to come by, but having now won it, the contingency approach seems so obviously correct that we are not likely to easily give it up."*

In short, I think it is time to change for change's sake. Not because I think I have the correct paradigm to replace open system models. But because I fear that some people have begun to treat contingency theory and other derivatives of open system modeling as the truth rather than as the most recent set of working assumptions. If we have begun to confuse the map with the territory, then it is time to change maps.

This concludes my litany of motives for abandoning open system models of organizations. In the next section I begin to outline some possible alternatives.

Some new directions

I have previously argued that although the *language* of Thompson's model is at the level of open systems, the actual *content* is wedged at Boulding's level three, the level of simple control systems; and most of the empirical research and analysis generated by the model has been at level one, the level of static frameworks. Therefore, one promising direction for empirical inquiry is actually to test Thompson's propositions at the proper level that reflects their dynamic rather than static content. One of the most sophisticated level two studies is Nystrom's (1975) analysis of the budgeting, workflow, and litigation processes within the Federal Trade Commission. Using 13 years of data from 1954 to 1971, Nystrom estimated a simultaneous, six equation model describing funds requested and appropriated, investigations completed, formal complaints, cease and desist orders, and litigations. Two of the equations included time-lagged variables, thus making the model (as well as the data) longitudinal or dynamic. Since the endogenous time lag was only one year in length, the model could not exhibit any natural cyclical behavior, but at least some dynamic characteristics were built into the model. Nystrom's strategy of analysis is important, and serves as a prototype of level two analysis.

Even better is the research of Hummon, *et al.* (1975). They have constructed a 'structural control' model of organizational change that is one of the few rigorous level three models in the field of organizational research. (The mathematics will be opaque to nearly all organization theorists, but is easily accessible to any undergraduate engineering major.) A structural control model presumes the existence of equilibrium points (not necessarily stable) within a system of variables, and a set of processes that describe how the system behaves when displaced away from those equilibrium points. If the system is stable, it will tend to converge on its equilibrium when displaced. Using Blau's (1970) model of structural differentiation as the content of the control model, and Meyer's (1972) data on governmental finance departments to test it, Hummon, *et al.* (1975), demonstrate the feasibility of estimating the equilibrium points, the control processes, and therefore the stability of the system. Just as Nystrom's (1975) research serves as a paradigm for level two modeling, the analysis of Hummon, *et al.*, (1975) provides a paradigm case for level three modeling.

If we are to go 'beyond open system models', we must first get there in content as well as in language. This suggests a second promising direction for inquiry, now primarily at the theoretical rather than empirical level. Before we can begin to answer questions about the behavior of open systems, we must first frame fruitful questions to ask. I believe that we have seriously misunderstood the nature of 'open systems', and have confused them with 'natural' or control systems. By an 'open system', we seem to have meant that the organization is influenced by the environment, or must take the environment into account, or can interact with the environment. But the interpretation advanced here is that a high variety environment is a *necessity* to an open system, not a problem, nor even a mere opportunity. The cognitive cycling produced by sensory deprivation provides an analog at the individual level of the phenomenon I have in mind. I am suggesting that there is a boundary between level three and level four systems across which the function of the environment undergoes a reversal. The human mind seems to be a system of sufficient complexity that it cannot continue to be a 'mind' in an environment of sensory deprivation. Those investigating the area of work motivation and job design have for some time realized the importance of task variety to continued satisfaction and productivity, especially for those with high growth needs (read 'high system complexity'?) (Hackman & Oldham, 1975). Is it unreasonable to conjecture that organizations of sufficient complexity also need high task variety in their environments? If so, what are the implications of Thompson's strategies of buffering, smoothing, standardizing, etc.? Do they constitute a self-imposed sensory deprivation for the organization?

If an organization is to advance across the boundary between a control system and an open system, it may need to be *flooded* with variety. Otherwise the control system will have time to develop buffers against a gradually developing complexity in the environment. A dunking in a sudden lack of structure is alleged to be what brings about change in sensitivity training groups. That insight suggests that the *rate* at which uncertainty overwhelms an organization will be more related to the complexity of its internal structure than just the *amount* of environmental uncertainty that happens to exist at the time of a cross-sectional study, or the pre-determined data collection periods of a longitudinal study. Since 'variety floods' cannot, by definition, be anticipated, an opportunistic research strategy is forced upon us if we wish to study the level three/level four metamorphosis. For example, we might wish to study organizations under conditions of natural disaster. In fact, Thompson (1967: 52-54) labels organizations that arise in response to disasters "synthetic organizations," and he attributes to them many open system characteristics quite different from the buffered systems operating under norms of rationality:

"...headquarters of the synthetic organization ... only occasionally emerge around previously designated officers... [A]uthority to coordinate the use of resources is attributed to - forced upon - the individual or group which by happenstance is at the crossroads of the two kinds of necessary information, resource availability and need... [W]hen normal organizations are immobilized or overtaxed by sudden disaster, the synthetic organization rapidly develops structure... [T]he synthetic organization emerges without the benefit of planning or blueprints, prior designations of authority, or formal authority to enforce its rules

		Experimen-talists	Theorists
Form of communica-tion	Verbal	66%	31%
	Publications	34%	69%
		100%	100%

Table 2

or decisions... [It has] great freedom to acquire and deploy resources, since the normal institutions of authority, property, and contract are not operating" (Thompson, 1967: 52-53).

In short, a synthetic organization is a self-organizing open system. But my only quibble with Thompson - a major one - is that such synthetic organizing processes are not limited to natural disasters and are far more common than he suggests.

To keep our models straight, we must be careful not to endow an open system with too many properties that characterize Boulding's higher levels of system complexity. For example, we should not attribute any desire or motivation or even tendency to the system to move from level three to level four, or to seek out environments rich enough in variety to maintain system means complexity, or to reproduce itself by means other than mitosis-like duplication, or to have a sense of self-awareness. Those are higher level properties. The sole property at issue in this immediate discussion has been an open system's capacity for self-organization and the important role of environmental variety in maintaining that capacity. Having established that caveat, we can move on to consideration of some of those higher level properties.

In previous sections I have already dealt, albeit briefly, with possible research questions about organizational birth and reproduction, and with phenomenological and socially constructed realities. But I have dealt only in passing with language and its relevance to organizational research. It is therefore to language that I should like to direct my attention here. Language plays at least four important and distinct roles in social behavior, including organizational behavior:

a. It controls our perceptions; it tends to filter out of conscious experience those events for which terms do not exist in the language;

b. It helps to define the meaning of our experiences by categorizing streams of events;

c. It influences the ease of communication; one cannot exchange ideas, information or meanings except as the language permits;

Silverman has addressed the first two of these functions in his action theory of organizations:

"Social reality is 'pre-defined' in the very language in which we are socialized. Language provides us with categories which define as well as distinguish our experiences. Language allows us to define the typical features of the social world and the typical acts of typical actors" (Silverman, 1971: 132).

In a sense, language is a technology for processing both information and meanings just as production technologies process inputs into outputs. Both types of technology constrain what inputs will be accepted and what transformations will be permitted. Languages vary in their capacity to process high variety information. For example, the language of written communication unaided by non-verbal cues is less able to detect complex events than is the verbal plus non-verbal language of face-to-face communication. Thus we would expect face-to face communication to be used more heavily in ill-structured fields such as 'general management' than in well-structured fields such as 'finance', with 'marketing' falling between them. Furthermore, in highly unstructured situations, even face-to-face communication may be inadequate for conveying the full meaning. We would therefore expect physical inspection to be most common in the poorly structured areas. This is precisely what Keegan (1974) found in a study of information sources used by headquarters executives of multinational corporations, as the following table (1) taken from Keegan's article shows:

		Field of Specialization		
		General Mgt.	Marketing	Finance
Type of information source	Documentary	18%	30%	56%
	Human (face-to-face)	71%	65%	44%
	Physical Inspection	11%	5%	0%
		100%	100%	100%

Table 1

Although Thompson ignores language as a variable of interest, an earlier classic in organization theory does not; in fact, March

Classic Complexity pp. 193-226

and Simon (1958: 161-169) make language a central feature of their analysis of communication in organizations. Like Silverman, they recognize the importance of language in perceiving and defining reality. But they offer a thorough (and largely ignored) treatment of the effects of language on the efficiency and accuracy of communication. They define language broadly to include engineering blueprints and accounting systems as well as 'natural' languages such as English. Standardized languages permit the communication of large amounts of information with minimal exchanges of symbols. On the other hand,

"...it is extremely difficult to communicate about intangible objects and nonstandardized objects. Hence, the heaviest burdens are placed on the communications system by the less structured aspects of the organization's tasks, particularly by activity directed toward the explanation of problems that are not yet well defined" (March & Simon, 1958: 164).

(But we should recognize the earlier point that objects become standardized by having terms in the language for referring to them. Objects are not standardized in and of themselves.)

For example, among physicists, experimental techniques and procedures probably are more ad hoc and nonstandardized than theories. Therefore, we would expect experimentalists to rely less on publications for obtaining research-relevant information from professional colleagues than theorists. Gaston (1972) in a study of particle physicists in the UK collected data to support that conjecture, as shown in Table 2.

With regard to the fourth function of language, the social influence function, Pondy (1974, 1976) has argued that possession of a common language facilitates the exercise of social control, and that organizations can be thought of as collections of 'jargon groups', within each of which specialized sub-languages grow up that set it apart from the other jargon groups in the organization. And the size and number of these jargon groups can be related to the age and size of the organization, its technology, and the rate of turnover of personnel (Pondy, 1975). Within a scientific community, the scientific paradigm provides a language for talking about professional matters. When this paradigm is poorly developed, as in academic departments of sociology, political science, and English, it has been shown that the turnover of department heads is more frequent than in departments with well-developed paradigms such as mathematics and engineering, the argument being that department heads in low paradigm fields are less able to exercise social control in the resolution of professional conflicts (Salancik, *et al.*, 1976)

Not all communication operates at the level of conscious, expressed language. Some recent papers have suggested that myths, stories, and metaphors provide powerful vehicles for exchanging and preserving rich sets of meaning (Milburn, 1975; Mitroff & Kilmann, 1976). This attention to the less conscious, less rational aspects or organizational language and communication provides one of the most exciting avenues for exploration open to us. It begins to approach the models characteristic of Boulding's level eight.

Let me try to place this brief discussion of the functions of language in organization theory in context. Imagine an organization of sub-humans incapable of the use of language. Modes of communication would be 'hard-wired' into the organisms. They would be incapable of reconceptualizing their relationships to each other, their technologies, or their environments. But language permits codification of those conceptualizations, and therefore sharing and social modification of them. Not only is language functional for the operation of the organization, but it is central to the evolution of organizational forms within the lifetimes of individual members. Mind need not wait for genetics to bring about change. If that premise is accepted, then the fundamental structures of language must be reflected in social organization. By 'fundamental structures' I mean such characteristics as the absence of the verb 'to be' in Turkish, Hopi, Hungarian and other languages, or the use of idiographic characters in Chinese. For example, it may be easier to communicate metaphorically in Chinese than in alphabet-based languages. And the fundamental structure of language may dwarf such surface characteristics as 'standardization' in their impact on organizational structure and behavior.

In a short paper, there is no easy way to conclude such an open-ended topic as attempting to define organization theory. I have barely scratched the surface of many intriguing ideas. I leave them knowingly but regretfully jagged, and turn in the last section to a discussion of a few implications of my position for the teaching and doing of organization theory.

Some implications

To discuss the implications for teaching and research of any theoretical position on management and organization theory is a tricky business. There is every likelihood that what we teach now to practitioners will create - assuming that they will practice what we preach - the very phenomena that we will have available to study in the future. Today's theories enact tomorrow's facts.

To deny this likelihood is to accept the ineffectiveness of our teaching; to admit it is to reject the role of scientist in favor of one closer to that of playwright. To be quite honest, I have been unable to resolve this paradox, and it circles buzzard-like over what I have to say in this concluding section.

The implications for research on organizations follow fairly directly from much of what I've already said, but the implications for teaching of practitioners will need to be made more actively explicit.

To summarize what I believe should now be obvious, if not agreeable, implications of my position for research:

a. Conceptually, the status of an organization shifts from that of an objective reality to one which in the extreme is phenomenologically represented in the subjective experience of individual participants, or more moderately is a socially constructed reality. Given such a conception, to endow such concepts as technology with measurable and perceivable attributes is questionable. Instead, we need to study how participants themselves come to invoke categories such as 'organization' and 'technology' as a means of making sense of their experience. The resulting meanings will frequently be 'stored' in organizational myths and metaphors to provide rationales for both membership and activity in organizations. The role that institutional leaders play in the creation of myths and metaphors is a worthwhile focus for study;

b. More generally, organizations are represented as collections of 'organizing rules' that generate observable behavior. While comparative analysis can document empirical regularities at the observable level, the true task of theory is to infer the generative mechanisms that produce the surface behavior *in each case*. That is, to develop a theory of the individual case is a meaningful scientific activity. Determining whether collections of individuals have the same theories is a proper task for comparative analysis. What I have in mind is analogous to discovering the relationship between a given acorn and oak, and subsequently establishing it for all acorn-oak pairs. (But that process will be intimately tied up with the definitions of 'acorn' and 'oak!') By implication, we must drop our reliance on comparative analysis as the only source of scientific generalizations about organizations;

c. These two conceptual hooks imply some radical methodological departures as well. I suspect that questionnaire design, large sample surveys, and multivariate analysis will need to recede in importance in favor of more abstract model-building (as in linguistics) and ethnographic techniques more suitable for studying meaning and belief systems. (The heavy investment in questionnaire design, large sample survey techniques, and multivariate analysis will, of course, occasion no resistance to these suggestions on the part of current practitioners of comparative analysis!) This is in no sense a suggestion that we return to the purely descriptive case study. Our aim is to find out how things work, and that can best be done one at a time. Whether a collection of individual cases work the same should be the end result of empirical inquiry, not the initial presumption as in comparative analysis. What is at issue is what we mean by the phrase, "how things work." Perhaps it would help to point out that the nature of causation changes as you ascend Boulding's hierarchy of complexity. Correlational models of causation implicit in comparative analysis are appropriate only at the levels of frameworks and clockworks, not at the level of blueprinted growth. And "how things work" is tied up with the nature of causation.

My attempt to summarize the implications for research (and the teaching of research) has turned out to be more abstract than I had hoped. Let me try to be a bit more specific in tracing some implications for teaching of practitioners.

Thompson's view of organizations suggests that administrators should be trained in the skills of 'co-aligning' environment, goals, technology, and structure in harmonious combination. And the conditions of harmony should derive from a rationality based on *organizational* well-being. These prescriptive out-takes from Thompson's descriptive analysis have, I believe, formed the primary basis for management training in organization theory for the past decade. The position I have advocated in this paper has a number of contrary implications for management education:

a. By highlighting the true open-system characteristics of organizations managers can perhaps be made aware of the environmental consequences of actions taken in the narrow interests of the organization, and shown the boomerang quality of organizational 'rationality' as the environment becomes more tightly coupled. Somehow, we need to generalize the concept of ecology and build it into the conscious calculus of administrative decision makers. My best guess is that the most effective - because experiential - way to do that is through large-scale, time-compressed simulations.

We may not be able to eliminate the motivation of self-interest, but we may be able to enlarge the manager's concept of self through such simulations.

b. By developing a typology of system dysfunctions and early warning signals, we may be able to train administrators to react adaptively when Thompson's harmony and co-alignment do not materialize according to plan. To my knowledge, nowhere do we now teach a diagnosis-and-treatment-of-macro-pathologies to managers or would-be managers.

c. I believe that the most radical implication of my position for management education derives from the view of organizations as language-using, sense-making cultures. In Thompson's view, the organization is an input-output machine, and the administrator is a technologist. In my view, the administrator's role shifts from technologist to linguist, from structural engineer to mythmaker. That is, a key function of management in a level 7 or 8 system, is that of helping the organization to make sense of its experiences so that it has a confident basis for future action. The administrator must have a skill in creating and using metaphors. This suggests the delightful conclusion that we should be teaching our institutional leaders not only capital budgeting and inventory control, but also poetry.

References

Alexander, H. G. (1967). *Language and thinking*, Princeton, CA: Van Nostrand.

Ashby, R. (1956). *An introduction to cybernetics*, London, UK: Chapman & Hall.

Bateson, G. (1972). *Steps to an ecology of mind*, New York, NY: Ballantine, ISBN 0226039056 (2000).

Beer, S. (1964). *Cybernetics and management*, New York, NY: Wiley, ISBN 0340045949 (1967).

Berger, P. L. and Luckman, T. (1966). *The social construction of reality*, Garden City, NY: Doubleday, ISBN 0385058985 (1967).

Blau, P. (1970). "A formal theory of differentiation in organizations," *American Sociological Review*, ISSN 0003-1224, 35: 201-218.

Boulding, K. (1968). "General systems theory: The skeleton of science," in Walter Buckley (ed.), *Modern systems research for the behavioral scientist*, Chicago, IL: Aldine, ISBN 0202300110, pp. 3-10.

Burns, T. and Stalker, G. M. (1961). The management of innovation, London, UK: Tavistock, ISBN 0198288786 (1994).

Chomsky, N. (1972). *Language and mind*, enlarged edition, New York, NY: Harcourt Brace Jovanovich, ISBN 052167493X (2006).

Cohen, M. D. and March, J. G. (1974). *Leadership and ambiguity*, New York,

NY: McGraw-Hill, ISBN 0070100632.

Crozier, M. (1964). *The bureaucratic phenomenon*, Chicago, IL: University of Chicago Press.

Cyert, R. M. and March, J. G. (1963). *A behavioral theory of the firm*, Englewood Cliffs, NJ: Prentice-Hall, ISBN 0631174516 (1992).

Etzioni, A. (1961). *A comparative analysis of complex organizations*, New York, NY: Free Press, ISBN 0029096200 (1975).

Gaston, J. (1972). "Communication and the reward system of science: A study of a national 'invisible college'," *The Sociological Review Monograph*, ISSN 0081-1769, 18: 25-41.

Geertz, C. (1973). *The interpretation of cultures*, New York, NY: Basic Books, ISBN 0465097197 (2000).

Goldsen, R. K. (1975). "The technological fix: Existentialist version," *Administrative Science Quarterly*, ISSN 0001-8392, 20: 464-468.

Haas, J. E. and Drabek, T. E. (1973). *Complex organizations: A sociological perspective*, New York, NY: Macmillan, ISBN 0023485507.

Hackman, R. and Oldham, G. (1975). "Development of the job diagnostic survey," *Journal of Applied Psychology*, ISSN 0021-9010, 60: 159-170.

Halberstam, D. (1972). *The best and the brightest*, New York, NY: Random House, ISBN 0449908704 (1993).

Hamilton, D. (1976). "A science of the singular?" CICRE, University of Illinois, School of Education, Urbana, Illinois, unpublished manuscript.

Harre, H. and Secord, P. F. (1973). *The explanation of social behavior*, Totowa, NJ: Littlefield, Adams & Co., ISBN 0822602695.

Hedberg, B. L. T., Nystrom, P. C. and Starbuck, W. H. (1976). "Camping on seesaws: Prescriptions for a self-designing organization," *Administrative Science Quarterly*, ISSN 0001-8392, 21: 41-65, also available at http://pages.stern.nyu.edu/~wstarbuc/Camping.htm.

Hummon, N. P., Doreian, P. and Teuter, K. (1975). "A structural control model of organizational change," *American Sociological Review*, ISSN 0003-1224, 40: 813-824.

Janis, I. (1972). *Victims of groupthink*, Boston, MA: Houghton Mifflin, ISBN 0395140447.

Keegan, W. J. (1974). "Multinational scanning: A study of information sources utilized by headquarters executives in multinational companies," *Administrative Science Quarterly*, ISSN 0001-8392, 19: 411-421.

Kimberly, J. R. (1976). "Contingencies in creating new institutions: An example from medical education," unpublished manuscript presented at the Joint EIASM-Dansk Management Center Research Seminar on *Entrepreneus and the Process of Institution Building*.

Lawrence, P., and Lorsch, J. (1967). *Organization and environment*, Cambridge, MA: Harvard University Press, ISBN 0875840647.

Leavitt, H. J. and Pondy, L. R. (eds.) (1964). *Readings in managerial psychology*, 1st edition, Chicago: University of Chicago Press.

March, J. G. and Simon, H. A. (1958). *Organizations*, New York, NY: Wiley, ISBN 063118631X (1993).

Milburn, T. W. (1975). "Metaphors as bases and results of organizational functioning," unpublished paper presented at the Illinois *Workshop on Radical Approaches to Organization Design*, Urbana, Illinois, Department of Business Administration.

Mitroff, I. and Kilmann, R. (1976). "On organizational stories: An approach to the design and analysis of organizations through myths and stories," in R. H. Kilmann, L. R. Pondy, and D. P. Slevin (eds.), *The management of organization design: Strategies and implementation*, New York, NY: American Elsevier, ISBN 0444001883.

Newell, A. and Simon, H. A. (1972). *Human problem solving*, Englewood Cliffs, NJ: Prentice-Hall, ISBN 0134454030.

Nystrom, P. C. (1975). "Input-output processes of the Federal Trade Commission," *Administrative Science Quarterly*, ISSN 0001-8392, 20: 104-113.

Perrow, C. (1967). "A framework for the comparative analysis of organizations," *American Sociological Review*, ISSN 0003-1224, 32: 194-208.

Pettigrew, A. M. (1976). "The creation of organizational cultures," unpublished manuscript, London Graduate School of Business Studies.

Pfeffer, J. (1976). "Beyond management and the worker: The institutional function of management," *Academy of Management Review*, ISSN 0363-7425, 1(2): 36-46.

Pfeffer, J. and Salancik, G. (1974). "Organizational decision making as a political process: The case of a university budget," *Administrative Science Quarterly*, ISSN 0001-8392, 19: 131-151.

Pondy, L. R. (1974). "The other hand clapping: An information processing approach to organizational power," unpublished paper presented at the Cornell University mini-conference on *Organizational Power*, School of Industrial and Labor Relations, December 6, 1976.

Pondy, L. R. (1975). "A minimum communication cost model of organizations: Derivation of Blau's laws of structural differentiation," unpublished manuscript, University of Illinois, Urbana.

Pondy, L. R. (1976). "Leadership is a language game," in M. McCall and M. Lombardo (eds.), *Leadership: Where else can we go?* Durham, NC: Duke University Press, ISBN 0822306271 (1984).

Pondy, L. R. and Boje, D. M. (1976). "Bringing mind back in: Paradigm development as a frontier problem in organization theory," Department of Business Administration, University of Illinois, unpublished manuscript.

Salancik, G. and Pfeffer, J. (1974). "The bases and uses of power in organizational making: The case of a university," *Administrative Science Quarterly*, ISSN 0001-8392, 19: 453-473.

Salancik, G., Staw, B. and Pondy, L. (1976). "Administrative turnover as a response to unmanaged organizational interdependence: The department head as a scapegoat," unpublished manuscript, Department of Business Administration, University of Illinois.

Schön, D. A. (1971). *Beyond the stable state*, New York, NY: W. W. Norton, ISBN 0393006859 (1973).

Schrödinger, E. (1968) "Order, disorder, and entropy," in W. E. Buckley (ed.), Modern systems research for the behavioral scientist, Chicago, IL: Aldine, ISBN 0202300110, pp 143-146.

Scott, W. R. (1976). "On the effectiveness of studies of organizational effectiveness," unpublished paper presented at the Workshop on Organizational Effectiveness, Carnegie-Mellon University, June 28-29.

Silverman, D. (1971). The theory of organizations, New York, NY: Basic Books, ISBN 0566055627 (1987).

Smith, R. A. (1963). Corporations in crisis, Garden City, NY: Doubleday, ASIN B0007DTD0S.

Starbuck, W. H. (ed.) (1971). Organizational growth and development, Baltimore: Penguin, ISBN 0140801367.

Staw, B. M. (1976). "Knee-deep in the big muddy: The effect of personal responsibility and decision consequences upon commitment to a previously chosen course of action," Organizational Behavior and Human Performance, ISSN 0030-5073, 16: 27-44.

Staw, B. M. and Fox, F. V. (1977). "Escalation: Some determinants of commitment to a previously chosen course of action," Human Relations, ISSN 0018-7267, 30(5): 431-450.

Staw, B. and Szwajkowski, E. (1975). "The scarcity-munificence component of organizational environments and the commission of illegal acts," Administration Science Quarterly, ISSN 0001-8392, 20: 345-354.

Thompson, J. D. (1967). Organizations in action, New York, NY: McGraw-Hill, ISBN 0765809915 (2003).

Tung-Sun, C. (1970). "A Chinese philosopher's theory of knowledge," in G. P. Stone and H. A. Farberman (eds.), Social psychology through symbolic interaction, Waltham, MA: Xerox College Publishing Co., ISBN 0471005754, pp. 121-140.

Weick, K. E. (1969). The social psychology of organizing, Reading, MA: Addison-Wesley, ISBN 0075548089.

Weick, K. E. (1976). "On repunctuating the problem of organizational effectiveness," unpublished paper presented at the Workshop on Organizational Effectiveness, Carnegie-Mellon University, June 28-29.

Whorf, B. L. (1956). Language, thought and reality, Cambridge, MA: MIT Press, ISBN 0262230038.

11. The functions of the executive Chapter 2: The individual and organization
Chester I. Barnard

Originally published in Barnard, C. I. (1938). *The Functions of the Executive*, ISBN 0674328035 (2005), pp. 8-21. The *E:CO* editorial team would like to thank Harvard University Press for their kind permission to reprint this chapter.

Introduction

It is something of a marvel to recognize just how much of what is only now coming forth concerning leadership by means of complex systems research was already anticipated in the carefully considered insights published by Chester Bernard as far back as 1938. In his book, *The Functions of the Executive*, Barnard described an integrated perspective placing the individual actions of leaders squarely within the context of a systemic understanding of organizations. In this regard, the economist John Kenneth Galbraith credited Barnard with had providing the most informative definition of organization of the time. Other researchers from Nobel Laureate Herbert Simon to James March have also been inspired by his insights - March suggesting that Barnard had actually instigated much of the later research and Simon pointing out the deep complexity perspective that Benard enunciated.

Chester Barnard (1886-1961) is believed to be the first executive who wondered out loud about what a business leader should be. Born in Malden, Massachusetts, his father was a mechanic; his mother died when he was five years old. At the age of fifteen, he worked as a piano tuner. He won an economics scholarship to Harvard and studied economics and languages. In the third year in Harvard, he dropped out of college due to a shortage of funds, but fortuitously landed a job that was to prove a great benefit to himself and American industry when he joined American Telephone and Telegraph to work as a statistician in 1909. He spent his entire working life with the company, becoming President of New Jersey Bell in 1927, finally retiring in 1952.

Chester Barnard's best-known book, *The Functions of the Executive* (1938), collected his eight lectures given at the Lowell Institute in Boston in 1937. Another well-known work is *Organization and Management* (1948). Barnard also made great contributions to not-for-profit foundations. He was a president of the United Service Organizations for National Defense (1942-1945) during World War II; state director of the New Jersey Relief Administration during the Depression; President of the Rockefeller Foundation and the General Education Board (1948-1952); Chairman of the National Science Board (1950-1956); and, Chairman of the National Science Foundation (1952-1954). In addition, Barnard was a co-author of the State Department report on International Control of Atomic Energy which went on to become a fundamental policy statement for the government in that area. Bernard was also known for his work supporting African American Soldiers of which he once said, "In the long run this accomplishment may be more important than anything we have done, for unity amid diversity is a fundamental problem of world peace" (Krass, 1998).

Overview of the excerpt "The individual and organization

We have chosen this specific excerpt because it highlights three key aspects of his approach that are of importance to complex systems research:

1. The duality of individuals as both autonomous agents and also actors influenced by their situation within a system;
2. Purpose as the underlying driver of cooperation among agents, and;
3. The distinction between effectiveness and efficiency in evaluating action and choices within a system.

When the individual is considered as an organizational actor, the relevant actions must be considered within an organizational purpose - purpose being a critical determinant of cooperation. It is not simply the capacity of individuals to act for their own benefit and to make choices toward that end which drives human events.

Just as importantly, it is the organization's capacity to limit the choices available to individuals that in the end channels coordinated action. If Chester Barnard were alive today, we believe he would be very excited about the insights being supplied into individual and systemic purpose through the study of complex systems. We suggest that a close reading of the excerpt juxtaposed with the findings of the papers in this special issue can lead to valuable cross-fertilizations between a visionary of the past and those of the present, and between leadership practice and leadership theory.

Jeffrey A. Goldstein
James K. Hazy

References

Crainer, S. (1998). "Chester Barnard," *The Ultimate Book of Business Gurus: 110 Thinkers Who Really Made a Difference,* Oxford, England: Captone Publishing Limited, ISBN 0814404480, pp. 11-14.

Wren, D. A. (2000). "Barnard, Chester Irving," *American National Biography Online,* www.anb.org.

Gabor, A. (2000). *The Capitalist Philosophers: The Geniuses of Modern Business - Their Lives, Times, and Ideas,* New York, NY: Times Books, ISBN 0812928202, pp. 67-82.

Krass, P. (1998). *Chester I. Barnard. The Book of Leadership Wisdom,* Canada: Wiley & Sons, Inc., ISBN 04712945501, pp. 37-42.

CHAPTER II

THE INDIVIDUAL AND ORGANIZATION

I HAVE found it impossible to go far in the study of organizations or of the behavior of people in relation to them without being confronted with a few questions which can be simply stated. For example: "What is an individual?" "What do we mean by a person?" "To what extent do people have a power of choice or free will?" The temptation is to avoid such difficult questions, leaving them to the philosophers and scientists who still debate them after centuries. It quickly appears, however, that even if we avoid answering such questions definitely, we cannot evade them. We answer them implicitly in whatever we say about human behavior; and, what is more important, all sorts of people, and especially leaders and executives, act on the basis of fundamental assumptions or attitudes regarding them, although these people are rarely conscious that they are doing so. For example, when we undertake to persuade others to do what we wish, we assume that they are able to decide whether they will or not. When we provide for education or training we assume that without them people cannot do certain things, that is, that their power of choice will be more limited. When we make rules, regulations, laws — which we deliberately do in great quantities — we assume generally that as respects their subject matter those affected by them are governed by forces outside themselves.

The significance of these observations may be made clearer by noting the extreme differences of conception regarding the "individual" — to take one word — in discussions of coöperation and of organizations and their functions. On the one hand, the discrete, particular, unique, singular individual person with a name, an address, a history, a reputation, has the attention.

THE INDIVIDUAL AND ORGANIZATION   9

On the other hand, when the attention transfers to the organization as a whole, or to remote parts of it, or to the integration of efforts accomplished by coördination, or to persons regarded in groups, then the individual loses his preëminence in the situation and something else, non-personal in character, is treated as dominant. If in such situations we ask "What is an individual?" "What is his nature?" "What is the character of his participation in this situation?" we find wide disagreement and uncertainty. Much of the conflict of dogmas and of stated interests to be observed in the political field — the catchwords are "individualism," "collectivism," "centralization," "laissez-faire," "socialism," "statism," "fascism," "liberty," "freedom," "regimentation," "discipline" — and some of the disorder in the industrial field, I think, result from inability either intuitively or by other processes to reconcile conceptions of the social and the personal positions of individuals in concrete situations.

These considerations suggest that in a broad inquiry into the nature of organizations and their functions, or in an effort to state the elements of the executive processes in organizations, a first step should be to set forth the position or understanding or postulates especially concerning the man, the "individual," and the "person," and related matters. Without such a preliminary survey it is quite certain that there will be unnecessary obscurity and unsuspected misunderstanding. This does not mean that I shall attempt either a philosophic or a scientific inquiry. It does mean that I must present a construction — a description or definite scheme — to which consistent reference is implied throughout this book.

Accordingly, in this chapter, I shall briefly discuss the following subjects: I, The status of individuals and the properties of persons generally; II, the method of treating individuals and persons in this book; III, certain characteristics of personal behavior outside coöperative systems; and IV, the meaning of "effectiveness" and "efficiency" in personal behavior.

10   COOPERATIVE SYSTEMS

I. Concerning, i, The Status of Individuals and
ii, The Properties of Persons

I

(*a*) First of all, we say that an individual human being is a
discrete, separate, physical thing. It is evident that every one
believes, or usually acts as if he did, in this individual physical
entity. For other and broader purposes, however, it seems clear
that no thing, including a human body, has individual independent existence. It is impossible to describe it, use it, locate
it, except in terms of the rest of the physical universe or some
larger isolation of it. For example, if the temperature of the
environment changes, that of the thing or of the body must
change (except within limits of adjustment biologically determined). Its weight is a function of gravitational attraction;
its structure depends upon gravity both directly and indirectly.
Thus at the outset we note that the human being, physically
regarded, may be treated either as an individual thing or as a
mere phase or functional presentation of universal physical
factors. Which is "correct" depends upon the purpose. When
the architect calculates the live-load capacities of his floor structures he is thinking of men not as individual human beings but
as functions of gravitational attraction; other aspects of them
he disregards.

(*b*) The mere body, however, whether for limited and practical purposes regarded as a physical object or regarded as a
phase or function of general physical factors, is not a human
being. As a living thing it possesses a power of adjustment, an
ability to maintain an internal balance, and a continuity, despite
incessant changes within and wide variations without itself.
Moreover, it possesses a capacity of experience, that is, an ability
to change the character of its adjustment as a result of its history. This means that the human body, viewed by itself, is an
organism, something whose components are both physical and

232

THE INDIVIDUAL AND ORGANIZATION   11

biological. Although the physical factors are distinguished from the biological they are not separable in specific organisms. In other words, living things are known by behavior, and all living behavior is a synthesis of both physical and biological factors. If either class of factors are removed, the specific behavior ceases to be manifest, and the physical form also undergoes changes that would otherwise not occur. But if a single organism is so composed, this means that it not only presents universal physical factors but also a long race history, so that the organism is an individual when we forget all these facts, and if we remember them becomes a whole conglomeration of things that we cannot even see.

(*c*) Human organisms do not function except in conjunction with other human organisms. This is true first because they are bisexual. It is also true because in infancy they require nurture. Moreover, the mere presence of organisms without reference to sexuality, parenthood, or infancy, compels interrelations or interactions between them; as physical objects they cannot occupy the same space, there is an interchange of radiant energy between them, they reflect light to each other. Biologically they compete for food — which is both a physical and a biological requirement.

The interactions between human organisms differ from those between mere physical objects or between a physical object and an organism in that experience and adaptability are *mutually* involved. The adaptation required and the experience relate not merely to things or functions determined by factors inherent in each of the organisms separately, but to the mutuality of reaction or adjustment itself. In other words, the mutual reaction between two human organisms is a series of responses to the *intention* and *meaning* of adaptable behavior. To the factors peculiar to this interaction we give the name "social factors," and we call the relationship "social relationship."

On first consideration, a physical thing endowed with life

that has interacted with other similar organisms becomes more and more unique, separate, distinct, just as a point where many lines cross seems to the mind more definitely a point than one where only two lines cross. But when we stop to think of the history of its physical components, of its long line of ancestors, and the extent to which it embodies the effects of actions of others, it becomes less and less distinct, less and less an individual, more and more a mere point where the crossing lines are more important than the place where they cross. The individual is then a symbol for one or more factors, depending on the breadth of our interest.

Sometimes in everyday work an individual is something absolutely unique, with a special history in every respect. This is usually the sense in which we regard ourselves, and so also our nearest relations, then our friends and associates, then those we occasionally meet, then those we know about, then those we hear about, then those that are in crowds, those represented by statistics, etc. And the farther we push away from ourselves the less the word "individual" means what it means when applied to you and me, and the more it means a spot showing some aspect of the world that commands our attention. Then an individual becomes not a particular man but only a worker, a citizen, an underprivileged man, a soldier, an official, a scientist, a doctor, a politician, an economic man, an executive, a member of an organization.

In this book we mean by the individual a single, unique, independent, isolated, whole thing, embodying innumerable forces and materials past and present which are physical, biological, and social factors. We shall usually not be concerned with how he came to be or why, except as this is directly involved in his relations to organization. When we wish to refer, as we frequently must, to aspects, phases, or functions of individuals, we shall use other words, such as "employee," "member," "contributor," "executive," or otherwise limit our reference.

THE INDIVIDUAL AND ORGANIZATION 13

II

The individual possesses certain properties which are comprehended in the word "person." Usually it will be most convenient if we use the noun "individual" to mean "*one* person" and reserve the adjectival form "personal" to indicate the emphasis on the properties. These are (*a*) activities or behavior, arising from (*b*) psychological factors, to which are added (*c*) the limited power of choice, which results in (*d*) purpose.

(*a*) An important characteristic of individuals is activity; and this in its gross and readily observed aspects is called behavior. Without it there is no individual person.

(*b*) The behavior of individuals we shall say are the result of psychological factors. The phrase "psychological factors" means the combination, resultants, or residues of the physical, biological, and social factors which have determined the history and the present state of the individual in relation to his present environment.

(*c*) Almost universally in practical affairs, and also for most scientific purposes, we grant to persons the power of choice, the capacity of determination, the possession of free will. By our ordinary behavior it is evident that nearly all of us believe in the power of choice as necessary to normal, sane conduct. Hence the idea of free will is inculcated in doctrines of personal responsibility, of moral responsibility, and of legal responsibility. This seems necessary to preserve a sense of personal integrity. It is an induction from experience that the destruction of the sense of personal integrity is the destruction of the power of adaptation, especially to the social aspects of living. We observe that persons who have no sense of ego, who are lacking in self-respect, who believe that what they do or think is unimportant, who have no initiative whatever, are problems, pathological cases, insane, not of this world, *unfitted for coöperation*.

14   COOPERATIVE SYSTEMS

This power of choice, however, is limited. This is necessarily true if what has already been stated is true, namely, that the individual is a region of activities which are the combined effect of physical, biological, and social factors. Free will is limited also, it appears, because the power of choice is paralyzed in human beings if the number of equal opportunities is large. This is an induction from experience. For example, a man set adrift while sleeping in a boat, awaking in a fog in the open sea, free to go in any direction, would be unable at once to choose a direction. Limitation of possibilities is necessary to choice. Finding a reason why something should *not* be done is a common method of deciding what should be done. The processes of decision as we shall see [1] are largely techniques for narrowing choice.

(*d*) The attempt to limit the conditions of choice, so that it is practicable to exercise the capacity of will, is called making or arriving at a "purpose." It is implied usually in the verbs "to try," "to attempt." In this book we are greatly concerned with purposes in relation to organized activities.

It is necessary to impress upon the reader the importance of this statement of the properties of persons. They are fundamental postulates of this book. It will be evident as we proceed, I think, that no construction of the theory of coöperative systems or of organizations, nor any significant interpretation of the behavior of organizations, executives, or others whose efforts are organized, can be made that is not based on *some* position as to the psychological forces of human behavior. There is scarcely a chapter in which those given are not exemplified.

An explicit statement on the question of free will, and certainly an avowed discussion of it, is usually to be found only in philosophic or scientific treatises. I must state my position because it determines the subsequent treatment in many ways.

[1] In Chapter XIV, "The Theory of Opportunism."

THE INDIVIDUAL AND ORGANIZATION   15

For this reason I should also add at this time that the exaggeration in some connections of the power and of the meaning of personal choice are vicious roots not merely of misunderstanding but of false and abortive effort. Often, as I see it, action is based on an assumption that individuals have a power of choice which is not, I think, present. Hence, the failure of individuals to conform is erroneously ascribed to deliberate opposition when they *cannot* conform. When the understanding is more nearly in accordance with the conception of the free will stated above, a part of the effort to determine individual behavior takes the form of altering the conditions of behavior, including a conditioning of the individual by training, by the inculcation of attitudes, by the construction of incentives. This constitutes a large part of the executive process, and is for the most part carried out on the basis of experience and intuition. Failure to recognize this position is among the important sources of error in executive work; it also results in disorganization and in abortive measures of reform, especially in the political field.

The narrow limitations within which choice is a possibility are those which are imposed jointly by physical, biological, and social factors. This is a conclusion from personal experience and direct observation of the behavior of others. It will be well illustrated in what follows. Hence, it may be true to say at the same time that a power of choice is always present and that the person is largely or chiefly a resultant of present and previous physical, biological, and social forces. Nor does this deny that the power of choice is of great importance. Though choice may be limited very narrowly at a given moment, the persistence of repeated choices in a given direction may ultimately greatly change the physical, biological, and social factors of human life. To me it is obvious it has done so.

16   COOPERATIVE SYSTEMS

II. The Treatment of Individuals and Persons in this Study

Conformably to what has been written above, persons are treated in two ways in this book. In this matter we also. conform to the ordinary practice of men as seen in their behavior. It is evident that we often regard men from the more nearly universal point of view, as phases, aspects, functions, which are greater spatially or durationally than individuals can be. For example, when we speak of managers, of employees, of voters, of politicians, of customers, etc., we have in mind certain *aspects* of individuals, certain kinds of activities of persons, not the whole individual. At other times we regard them as specific objective entities in discriminating their functional relationship to us or to other larger systems. We take into account the whole of the individual so far as we can. In practice we shift from one point of view to the other, or through rather vague intermediate positions, depending upon the circumstances and our purposes, with amazing skill in some cases; but we become much confused in talking about which point of view applies to our statement. In the latter respect only am I now trying to differ from usual practice.

In this book persons *as participants in specific coöperative systems* are regarded in their purely functional aspects, as phases of coöperation. Their efforts are de-personalized, or, conversely, are socialized, so far as these efforts are coöperative. This will be more specifically justified as a method of procedure in Chapter VI. Second, as *outside* any specific organization, a person is regarded as a unique individualization of physical, biological, and social factors, possessing in limited degree a power of choice. These two aspects are not alternative in time; that is, an individual is not regarded as a function at one time, as a person at another. Rather they are alternative aspects which may simultaneously be present. *Both are always present in coöperative systems*. The selection of one or the other of these

238

THE INDIVIDUAL AND ORGANIZATION   17

aspects is determined by the field of inquiry. When we are considering coöperation as a functioning system of activities of two or more persons, the functional or processive aspect of the person is relevant. When we are considering the person as the *object* of the coöperative functions or process, the second aspect, that of individualization, is most convenient.

As to any given specific coöperative system at a given time, most individuals in a society have no functional relationship in any direct sense.[2] Individuals connected with any given coöperative system have a dual relationship with it — the functional or internal relationship which may be more or less intermittent; and the individual or external relationship which is continuous, not intermittent. In the first aspect, some of the activities of the person are merely a part of a non-personal system of activities; in the second aspect the individual is outside, isolated from, or opposed to the coöperative system. It will be made evident as we proceed that in practical operation as well as in our analysis this dual-aspectual treatment of the individual is recognized and required.

III. THE BEHAVIOR OF INDIVIDUALS

It will be useful at this place further to consider persons in their individual aspect, as external to any coöperative systems. In this aspect persons choose whether or not they will enter into a specific coöperative system. This choice will be made on the basis of (1) purposes, desires, impulses of the moment, and (2) the alternatives external to the individual recognized by him as available. *Organization results from the modification of the action of the individual through control of or influence upon one of these categories.* Deliberate conscious and specialized control of them is the essence of the executive functions.

We shall call desires, impulses, wants, by the name "motives." They are chiefly resultants of forces in the physical, biological,

[2] See p. 84.

18   COOPERATIVE SYSTEMS

and social environments present and past. In other words, "motives" are constructions for the psychological factors of individuals in the sense previously discussed in this chapter, inferred from action, that is, after the fact. No doubt sometimes what we mean by "imagination" is a factor in the present situation. No doubt also, persons can occasionally be aware of their "motives." But usually what a man wants can be known even to himself only from what he does or tries to do, given an opportunity for selective action.[3]

Motives are usually described in terms of the end sought. If the action of a man is to obtain an apple, we say the motive of the action is to obtain an apple. This is misleading. The motive is rather the satisfaction of a "tension" resulting from various forces; and often we recognize this. We say that "the motive is to satisfy hunger." This is often surmise. The motive may be purely social — to give the apple away; or it may be a means of social action to buy something else. In most cases the end sought or the action taken represents motives of composite origin — social and physiological. This cannot be determined, and is usually unknown to the person whose action is involved. That the motive is of complex origin, however, is often clearly indicated by the importance of the conditions or circumstances which surround the end consciously sought. For example, a boy wants an apple, but it is evident that he wants an apple on the farmer's tree — not one at home or in the store. It is common experience that specific objects sought are often only sought under certain conditions or by certain processes. Of this we may not be aware unless the objects are offered or are available under other conditions or other processes which we reject.

[3] I do not necessarily mean by this that in any *specific* situation the motives of men may usually be determined by what they do or say at that time. On the contrary there are many situations in which the motives of men are to be only inferred by (1) what they do and say in this situation; (2) what they have said and done in the past in similar and in dissimilar situations; (3) what they do and say after the situation.

THE INDIVIDUAL AND ORGANIZATION 19

The activities incited by desires, impulses, wants — motives — sometimes result in the attainment of the end sought and satisfaction of the tension. Sometimes they do not so result. But they always have other effects which are not sought. Usually these unsought effects are regarded as incidental, inconsequential and trivial. For example, a man running to catch an animal for food gives off heat energy to the atmosphere, pulverizes a small amount of gravel, tears off a bit of skin, and somewhat increases his need of food while attempting to secure it. At other times consequences not sought for are not regarded as trivial. For example, the man running may move a stone which starts an avalanche which destroys his family, or his dwelling, or his stock of stored food.

IV. EFFECTIVENESS AND EFFICIENCY IN PERSONAL BEHAVIOR

The statement in the preceding paragraph is one of facts so obvious that they are neglected. They are among those of first importance in this study. They lead to distinctions in the meanings of the words "effective" and "efficient" both in relation to personal action and to organization action. We shall at this time consider their significance only as respects personal action.

When a specific desired end is attained we shall say that the action is "effective." When the unsought consequences of the action are more important than the attainment of the desired end and are dissatisfactory, effective action, we shall say, is "inefficient." When the unsought consequences are unimportant or trivial, the action is "efficient." Moreover, it sometimes happens that the end sought is not attained, but the unsought consequences satisfy desires or motives not the "cause" of the action. We shall then regard such action as efficient but not effective. In retrospect the action in this case is justified not

20   COOPERATIVE SYSTEMS

by the results sought but by those not sought. These observations are matters of common personal experience.

Accordingly we shall say that an action is effective if it accomplishes its specific objective aim. We shall also say it is efficient if it satisfies the motives of that aim, whether it is effective or not, and the process does not create offsetting dissatisfactions. We shall say that an action is inefficient if the motives are not satisfied, or offsetting dissatisfactions are incurred, even if it is effective. This often occurs; we find we do not want what we thought we wanted.

The specific ends sought by men are of two kinds, physical and social. Physical ends are material objects and physical conditions,[4] such as warm air, light, shade, etc. They are found in a "purely" physical environment. They are often found also in conjunction with a social environment. Social ends are contact, interrelations, communication, with other men. Such ends usually must be sought in a general social environment and always in some specific physical environment. Hence, whatever the specific ends, they serve to satisfy complex motives of persons. Usually a specific end of a physical class involves social consequences not sought. Always a social end involves physical consequences not sought.

The actions through which ends are sought are always physical (or physiological); they may also be social. In either case they involve unsought consequences that may give satisfaction or dissatisfaction. Social processes are those in which the action is a part of the system of actions of two or more men. Its most common form is verbal communication.

In accordance with the above we may say that the motives of men who enter into coöperation are almost invariably composite physiological and social motives in the sense that they are at least socially and physiologically conditioned. They may

[4] Including living things. Usually in this book we do not discriminate between biological and physical factors except with reference to the human being.

THE INDIVIDUAL AND ORGANIZATION   21

be predominantly physiological or predominantly social, but usually neither influence may be safely regarded as predominant.

In this chapter I have attempted to state the position essential to the conceptual scheme developed in this book as the means of presenting a theory of organization and a significant description of the executive processes. On the one side, those philosophies that explain human conduct as a presentation of universal forces, that regard the individual as merely responsive, that deny freedom of choice or of will, that make of organization and socialism the basic position, are found to rest upon facts that are widely observed and that govern men's behavior and thought in social situations. On the other side, those philosophies that grant freedom of choice and of will, that make of the individual an independent entity, that depress the physical and social environment to a secondary and accessory condition, are also consistent with other facts of behavior and thought. I undertake no reconciliation of the opposition in these philosophies or whatever scientific theories they may rest upon. For the present, at least, the development of a convenient and useful theory of coöperative systems and of organization, and an effective understanding of the executive processes, require the acceptance of both positions as describing aspects of social phenomena. What, then, is needed for our purposes is to state under what conditions, in what connections, or for what purposes one or the other of these positions may be adopted usefully, and to show how they may be regarded as simultaneously applicable. Coöperation and organization as they are observed and experienced are concrete syntheses of opposed facts, and of opposed thought and emotions of human beings. It is precisely the function of the executive to facilitate the synthesis in concrete action of contradictory forces, to reconcile conflicting forces, instincts, interests, conditions, positions, and ideals.

www.ingramcontent.com/pod-product-compliance
Lightning Source LLC
Chambersburg PA
CBHW070447100426
42812CB00004B/1227